OKLAHOMA GEOLOGICAL SURVEY.

CIRCULAR NO. 6.

ANIMAL AND PLANT LIFE IN OKLAHOMA

BY

OKLAHOMA GEOLOGICAL SURVEY

NORMAN.

October, 1917.

ANIMAL AND PLANT LIFE IN OKLAHOMA.

CONTENTS.

Page

Introduction 5

Chapter 1.—Animal life in Oklahoma 7

Introduction 7

Hoofed animals 8

Buffaloes 8

Antelope 8

Deer 8

Flesh-eating animals 9

Rocky Mountain lions 9

Bears 9

Loafer wolf 9

Coyotes 10

Badgers and skunks 10

Other flesh-eating animals 11

Gnawing animals 11

Beavers 11

Prairie dogs 11

Gophers 12

Rabbits 13

Small rodents 13

Moles 14

Bats 14

Birds 15

General statement 15

Wild turkeys 15

Prairie chickens 15

Quail 16

Hawks and owls 16

Woodpeckers 17

Eagles and vultures 17

Water and shore birds 17

Crows 18

English sparrows 18

Song birds 18

Fish and mussels 20

Fishes 20

Mussels 20

Page

Reptiles 21
    Snakes 21
    Turtles 22
    Lizards 22
Insects 22
Summary 23
Classified list of wild animals in Oklahoma 24
Chapter II.—Plant life in Oklahoma 37
Introduction 37
Physiography and geology 37
    Salt Plains 38
    Gypsum Hills 39
    Sand Hills region 40
    Black Mesa region 41
Soils 41
Precipitation 42
Life zones 42
Relation of physical conditions to the flora 42
Relation of our flora to that of adjacent regions 43
Distributing agencies 44
Plant families in Oklahoma 46
Trees and shrubs 55
A list of trees and shrubs in Oklahoma 58

ANIMAL AND PLANT LIFE IN OKLAHOMA.

INTRODUCTION.

The urgent needs of the public schools in Oklahoma for available information concerning the animal and plant life, and similar demands of the general public for the same information, lead to the preparation of the present publication. In connection with the regular duties of the field men of the Oklahoma Geological Survey, a large amount of available information is gained concerning the natural resources of the State, which is of value from an educational and scientific view, as well as from an economic standpoint. The cooperation of the members of the Oklahoma Academy of Science has made such a publication possible. These chapters on the animal and plant life of Oklahoma were published as Chapters VIII and IX in Bulletin No. 27, Geography of Oklahoma, recently issued by the Oklahoma Geological Survey.

The chapter on animal life in Oklahoma was written by Mr. Howard Cross, of the Department of Zoology, Oklahoma University. Dr. H. H. Lane, Head of the Department of Zoology, offerede much valuable assistance in the preparation of this chapter, and checked the scientific and common names used in the list of wild animals. Mr. Frank Rush, Forester in charge of the United States Forest and Game Preserve, in the Wichita Mountains, also furnished much information of value concerning the animal life of the State. Credit is due to many others who contributed to the data used in the chapter.

The chapter on plant life in Oklahoma was compiled from the manuscripts of unpublished reports on the plant life of Oklahoma, by Mr. G. W. Stevens, formerly Head of the Department of Biology of Northwestern State Normal, Alva, and Mr. C. W. Shannon, Director of the Oklahoma Geological Survey. These manuscripts mentioned include a report on the flowering plants of Oklahoma, by Mr. Stevens, and a report on the trees and shrubs of Oklahoma by Mr. Shannon.

C. W. SHANNON, Director.

OKLAHOMA GEOLOGICAL SURVEY,
Norman, October 1, 1917.

A. HEAD OF BEAVER DAM ON WOLF CREEK, HUNTER'S RANCH, NEAR FARGO, ELLIS COUNTY.

B. SHOWING TREE CUT BY BEAVER AT ABOVE LOCALITY.

CHAPTER I.

ANIMAL LIFE IN OKLAHOMA.*

INTRODUCTION.

Oklahoma at one time had a wealth of animal life that seemed almost impossible of exhaustion. Not only were birds and animals found here in unbelievable numbers, but the range in species was perhaps greater than in any other area of like size in the western part of the United States. Many of the animals characteristic of the mountainous regions of the west extended to the foot hills of the Rockies and often occurred in the western part of this State. Animals commonly found in the rough and timbered regions of the east came as far westward as the Ozark Mountains and remained in Oklahoma in considerable numbers, while the roving herds of the vast open plains of the northwest found ample pasturage and shelter here during the winter months and sometimes grazed beyond the southern boundaries of the State. To this immense throng must be added the myriads of semi-migratory birds and animals of the south that were found during a part or all of the year within the present limits of Oklahoma. The location of the State makes it a sort of meeting place, or focal point, for a variety and abundance of animal life that would not be possible under any other conditions.

The multiplicity of habitats existing in Oklahoma also makes possible a range in animal life that few states enjoy. The arid regions of the western part of Oklahoma have, or had, a large representation of all the varied fauna characteristic of the great semi-desert region to the north and west, while the mountains within the State offer range and protection to a vast assemblage of animals that our rapidly advancing civilization has not yet been able to destroy. Our broad streams and lakes have furnished shelter and food to water and shore animals that may still be found in abundance in regions not frequented by man, and the timbered regions of the south and east support a wealth of woodland forms that could not exist in localities where forests do not abound.

The changes incident to the settlement of a new country by man unavoidably cause the destruction of the feeding grounds and shelter of the wild animals, and as the improvement is pushed farther and farther

*This chapter was prepared by Howard Cross, Department of Zoology, Oklahoma University.

into the wild retreats, the native animals become less and less abundant and finally must perish altogether unless given some measure of protection. The advent of man into Oklahoma has, of course, very materially lessened many kinds of wild life, but fortunately only a few species have been lost entirely, and the location of a National Game Preserve within the State, and the provision recently made for the establishing of State game preserve, give assurance that no species now living in Oklahoma will ever be reduced to the point of extinction. We shall always have at least a remnant or sample of the faunal wealth that once abounded in this State.

HOOFED ANIMALS.

BUFFALOES.

The buffalo is the largest wild animal that has lived in Oklahoma in modern times. Within the memory of men now living, this noble animal roamed over the prairies in countless numbers, and at times whole valleys were covered with vast, unbroken herds. Great as this throng was, it went so quickly after the coming of the railroads that men could not believe it gone. In the fall of 1884 the hunters came to kill the buffaloes for their hides, which they sold for one dollar and a half apiece. but none could be found. "They are gone north," the gunners said, but they had not gone north; they had all but gone to utter extinction! At one time there were only twelve buffaloes in the State, but now there are eighty-four head in the Wichita National Game Preserve in Comanche County, and a small herd at the 101 Ranch in Kay County. The splendor and vigor of this remnant, however, are gone and they trail after a wagon to be fed, as broken and fearless as domesticated cattle.

ANTELOPE.

Antelope, while never so abundant as the buffalo, were very numerous in this State before the coming of the cowboy and his civilization. They often grazed in herds of several hundred and were well adapted to live in rough country. They remained somewhat longer than the buffalo, for the last antelope in Oklahoma, which lived in the southern part of Ellis County, were not killed until 1910. Since that time not a free antelope has lived within the State, and only a single specimen is held in captivity. This is a male and is on the Government Preserve in the Wichita Mountains.

DEER.

Deer were one time very abundant in the timbered sections of this State, and on account of their wonderful speed and cunning they have evaded the destructive forces of man and may yet be found running wild. Their range is now restricted to the mountainous regions, and it is doubtful if there are more than one hundred and fifty wild deer in the State today. Most of these are in the Kiamichi Mountains, where the persistent and criminal efforts of hunters to kill them will no doubt soon be successful in securing the last specimen.

FLESH-EATING ANIMALS.

ROCKY MOUNTAIN LIONS.

The larger flesh-eating animals were never numerous here, but there were enough during the time of early settlement to cause considerable damage to stock and to instill a constant fear in the lonely traveler. The Rocky Mountain lion was known in this State as late as 1900, and it is possible that a few stragglers are yet in some of the mountain districts, but they are difficult to find and are not often seen. One came as far east as Grant County in 1895. This animal followed Pond Creek with its sheltering timber, and left the protection of the woods at Jefferson long enough to spring into a man's yard and devour a pig. A few nights after the occurrence just narrated, a farmer living west of Jefferson was annoyed by the low, frightened growl of the family dog. When the door was opened the dog shot out into the darkness and a sharp yelp was heard. The next morning the hind leg of this courageous but imprudent dog was found about one hundred yards from the house. This is the last record of the misdeeds of this Rocky Mountain lion before it was finally hunted down and killed.

BEARS.

Bears are in many ways the most interesting animals of this State. They have never been abundant and have always been restricted in range to the more inaccessible timbered regions of the mountains. These animals have inflicted very little damage on the stock raising interests of the State, and have never been known to attack man unless driven into a corner, or wounded. Bears, as is the custom of all such animals, go into their dens early in December and sleep through the cold winter months, and emerge with the coming of warm weather in the spring. They spend much time at play when in the company of other bears, and if taken when young make very amiable pets. Their thick heavy coats protect them from injury even in very rough frolics, and they often acquire habits of play that make it imprudent to present the bare hand to even the most gentle of bruin pets. They feed on succulent roots, insect larvae, honey, fish, frogs, and almost any other kind of flesh they can secure. At one time it was thought that two kinds of bears lived in Oklahoma, but the small Black bear is probably the only specie which is found here. In the early spring when it emerges from the long winter's sleep, its usual coat of black has become very long and has a dusky brown color that is perhaps responsible for mistaking it for the so-called Brown bear.

LOAFER WOLF.

Wholesale destruction marks the path of the large Gray or Loafer wolf throughout its range. This animal was never abundant enough to extend over any considerable portion of the State, but in bands of from four to eight they were able to kill deer and buffalo, and with the coming of cattle and sheep they were responsible for a destruction that aroused a widespread demand for their speedy extermination. The "six-shooter," so characteristic of the western cowboy, was originally carried for the purpose of killing this wolf, but the practice soon spread far beyond the limits of the Loafer range and probably will be kept alive forever in the "movies." The ranch owner furnished the guns and ammunition to the cowboys and they were given the skin as a special reward if they were fortunate enough to kill one of these destructive beasts. The Loafer wolf generally frequents the rough country and forages over definite ranges which he covers in cycles every seven days. Settlers have taken advantage of this unusual habit and have proceeded by systematic poisoning to protect their herds from these marauders. The Loafer is now practically extinct in this State, and its passing will be a signal for unbounded rejoicing by cattle men everywhere who have herds within its range.

COYOTES.

The coyote is one of those unfortunate members of our animal society whose position has not been understood and whose services have never met with the appreciation which they deserve. These animals are distributed throughout the State but do not occur in large numbers in any particular locality. The coyote has long been known as a chicken thief, and does occasionally frequent poultry yards that have been carelessly left open at night, but it renders a service to the farmer in the destruction of large numbers of mice, rats, gophers, and rabbits that pays many times over for its one unfortunate indulgence. The coyote may not be an altogether desirable resident of a crowded agricultural district, but it is capable of much good and by no means deserves the unenviable reputation which it has.

BADGERS AND SKUNKS.

With the possible exception of hawks and coyotes no living animals have been more unjustly persecuted than the badger and the common skunk. Both have a few bad habits, but if a ledger of all their services were kept it would show such a balance in their favor as would put to shame the army of thoughtless men and boys who annually pursue to certain and unmerciful death these valuable allies of the farmer.

Badgers were at one time fairly numerous in this State, but persistent and systematic killing by everybody has so reduced them that only a few remain. They can live in arid regions where the annual rainfall is so slight that it would seem impossible for any animal to survive. They live in burrows, and have flat, compact bodies with powerful fore feet and claws which enable them to dig with a rapidity that few animals attain. They are thus equipped to dig out and destroy myriads of ground squirrels, gophers, and prairie dogs, which, on account of their habits of life, are practically immune from the attacks of all other predatory animals.

The skunk, of which several kinds occur in Oklahoma, is scattered more generally throughout the State and perhaps has a few more bad habits than the badger, but the record in its favor is scarcely less astounding. Skunks are preeminently animals of the field and wood, and their occasional appearance in the farm yard is accidental. They work, for the most part, at night, and the number of noxious insects, mice, and rats which they annually unearth and devour is almost beyond belief. The furs of these animals are becoming very valuable and a splendid income awaits the boy who has the ingenuity and courage to collect a number of skunks and organize the first fur-growing industry in Oklahoma.

OTHER FLESH-EATING ANIMALS.

Other flesh-eating animals existing in Oklahoma are the bobcat, raccoon, o'possum, weasel, and mink. These animals exist abundantly in some portions of the State, while in others they have already been exterminated. The weasel and bobcat were once found on the open plains wherever prairie dogs lived, but they have long since been driven to the protection of the timber, and even here they are only rarely seen.

GNAWING ANIMALS.

BEAVERS.

The beaver is the largest gnawing animal in Oklahoma, and easily leads the mammals of the world in mechanical and engineering skill. Beavers were at one time common in many of our smaller streams, and the line of their retreat is marked by dams and houses which they were forced to abandon hurriedly, in their efforts to keep away from the habitations of man. Wherever the banks were too low for the beavers to burrow into, they constructed dams across the channels and often raised the water as much as four or five feet, forming large artificial ponds with a depth sufficient to protect the entrance to their houses, and to provide for storing a food supply large enough to carry the beaver colony through the period when the ponds were frozen over and they were cut off from the usual food beyond the limits of the water's edge. The trees generally selected by the beaver for food include the birch, cottonwood, hickory, ash, and willow. To secure the tender limbs and younger shoots they often "cut" down trees as large as a foot in diameter. If only the smaller limbs were used they were cut off and dragged to the pond, but if the trunk was needed for strengthening the dam, a channel was dug from the pond to the tree and the timber was floated into position, as is the custom of lumbermen in all parts of the world. There is only one colony of these interesting animals now within the State. This is located on a small stream in Ellis County. It is hoped that the men who own the adjoining land will see that measures are taken to prevent this last picturesque beaver "town" from being depleted of its industrious population.

PRAIRIE DOGS.

Prairie dogs, which exist so abundantly in many parts of Oklahoma, are social creatures and live together in communities called towns. A colony may cover only a part of an acre, or it may be miles in extent. The population ranges accordingly from a few individuals to as many as 100,000 in the larger towns. They burrow into the ground to a depth of about 12 feet and they extend the den in a horizontal direction for a distance of from 10 to 15 feet. Any number of lateral channels may extend from this main run. Usually only a single pair occupies a burrow, except in the spring when the four young arrive. The ground for a considerable distance around the mouth of the burrow is stripped free of vegetation, and in thickly settled communities these barren spaces connect, so that large areas of grazing land are sometimes entirely cleared of grass by the presence of these animals. Prairie dogs are not really dogs in any sense of the word, but are large squirrels, and as they subsist entirely on roots, grasses, grain, and vegetables there is no reason, except perhaps the name, why they should not be as palatable as timber squirrels or rabbits. They are preyed upon by wolves, coyotes, bobcats, weasels, and the larger hawks, but since man has driven away or destroyed these natural checks the prairie dogs have increased at a corresponding rate and artificial control must be put into effect. The United States Biological Survey has perfected an effective and inexpensive method for poisoning the prairie dog, and its general use throughout the State is bringing back into cultivation large areas that were formerly occupied by these animals.

GOPHERS.

The common pocket gopher is by all odds the most destructive animal in Oklahoma. We have at least two species in this State; a small one found widely distributed but never numerous; and the large red pocket gopher that is causing more waste at the present time than any other living creature within the State. The gopher is the builder of the countless mounds of dark earth that we see thrown up over prairies and fields wherever we go. Some alfalfa patches have been more than one-tenth covered by these tell-tale hills. The gopher has a compact, muscular body, and with its powerful front feet and claws is able to burrow rapidly through the soft earth. The burrow or "run" usually extends about a foot beneath the surface, and at short intervals vertical tunnels extend upward to the top of the ground and through these the excavated soil is carried out and thrown up in small heaps. A single gopher occupies a burrow and often extends it several hundred feet in a season. The animals live, for the most part, on the roots and tubers which they find in their journeys through the ground. The gopher sometimes emerges from its burrow and fills its pockets with tender leaves and grain, which it always carries into the burrow before eating. Stock are often injured by stepping into gopher burrows, and many a cowboy's life has been lost in a fall caused by his mount plunging a leg into one of these burrows. A few kernels of poisoned grain placed in each burrow will rid a farm of these destructive pests within a reasonably short time. Little, however, will be accomplished unless whole communities cooperate, for a single farm cleared would soon be overrun from adjoining territory. The animals are still very numerous in certain parts of the State, and unless some concerted action is taken to check them in the localities where they abound, they will presently be doubling the damage they are now inflicting. The gopher rarely is seen above the ground and as a result has few natural enemies except the badger and the bull-snake, both of which have been so reduced by man that now the gopher is practically free to multiply unmolested. The responsibility rests entirely with the land owner; he must act now or pay a constantly increasing penalty.

RABBITS.

The most frequently seen and best known mammals of Oklahoma are rabbits. They refuse to be exterminated and are found in every locality in the State. We have two kinds of rabbits—the large, long-legged variety with athletic form, known as the jack rabbit, and the much smaller rabbit, known as the cottontail. The jack rabbit frequents the open plains. It does not burrow, but depends on its keen vision and powerful legs to keep it out of danger. The nest is made in a small, obscure depression in the ground and lined with soft wool taken from the breast of the mother. The cottontail frequents the wooded and more protected areas. It has short legs, wholly unfit for prolonged flight, but easily makes up for this loss in swiftness and cunning. It burrows well under stones or projecting roots, where it is free from the usual annoyance of men and dogs. Then, too, its color so harmonizes with its surroundings that it is often obliged to seek flight to avoid being trodden upon. Both jack rabbits and cottontails are wonderfully prolific, often rearing as many as four litters during the summer. They feed on grasses, grain, growing wheat, vegetables, fruit, and in winter, when pressed by hunger, they sometimes gnaw the tender bark from young trees. A few wisps of green alfalfa hay scattered regularly about the orchard in winter is the best protection against such attacks. The wholesale destruction of hawks, owls, wolves, and coyotes has removed the natural checks on rabbit production and occasionally they become so numerous in certain localities as to inflict great damage on the growing crops and trees. If properly cared for, their bodies can be made an important contribution to the food supply of the State. The men who hunt have done much to hold the rabbits in check, but sometimes this is not sufficient to keep them within the proper limits, and large hunts organized on a competitive basis often result in the slaughter of thousands in a single day. These hunts sometimes embrace as much as one-third of a county, and the men and boys taking part number into the hundreds.

SMALL RODENTS.

In addition to the gnawing animals discussed above, Oklahoma has the muskrat and the usual army of small rodents that exist abundantly throughout the entire State. The red and gray squirrels are found wherever timber conditions make it possible for them to live, and in the southern part of the State there are two species of "flying" squirrels. The striped ground squirrel, smaller than its timber cousins, burrows into the ground in meadows and grain fields and is responsible for an annual damage of more than $100,000. Rats and mice exist in abundance throughout the State, and the destruction which they cause in fields and store-houses is common knowledge. But we can avoid much of this loss. If shelter such as heavy grass corners, weeded fence rows, mulching, and unraked hay, is burned or removed, and skunks, hawks, owls, weasels, and bull-snakes are not foolishly killed on account of superstition and hearsay, but are protected as friends, the army of these destructive rodents will soon be so lessened that the annual damage from this source will be reduced to very small proportions.

MOLES.

Moles have long been the unhappy victims of circumstantial evidence. They spend most of their active life in plowing through the soft earth in search of insects, and are guilty of only a few of the crimes with which they are so often charged and convicted. Their food consists almost entirely of insects, and if they are confined in a cage with nothing but vegetable food they soon starve. Moles are attracted to localities where burrowing insects are most numerous, and their presence in a yard or garden indicates that unseen insect pests are at work beneath the ground. The farmer observes a fresh mole run extending the length of a row of planted seeds, and if the seeds fail to come up the damage is charged to the mole and every effort is made to see that summary justice is executed. In reality, the mole was attracted by the grubs that were feeding on the sprouting roots, and was not even remotely interested in the seed. Then to further cloud the real issue, field mice take advantage of the protecting shelter of the mole tunnel and follow up, destroying all seeds and roots that have been exposed by the mole. To these trespassers is directly chargeable most of the destruction usually attributed to the mole. However, the mole is not entirely blameless. Its runs, when extended into lawns, render them unsightly, and many small roots are displaced and plants killed by the tunneling of this industrious animal. They should of course be kept out of yards and gardens, and this can be done by persistently trampling down their runs. The moles will thus be saved for carrying on the work of destroying underground insects in the fields, for which nature has so admirably fashioned them.

BATS.

Bats perform much the same service in the air that moles do beneath the ground. Their diet is made up largely of insects, and they contribute in no small way to holding these pests in check. It is true that we do not often see them at work, for they hang head downward under some protecting object during the day, but at dusk they come from their hiding and with a wierd, broken flight pursue and capture almost every kind of insect that moves about after night. Bats have little to commend them in the way of beauty or interest, but their persistent destruction of insects represents a saving to the State that should never be overlooked.

BIRDS.

GENERAL STATEMENT.

Oklahoma has a wealth of bird life of which she may be justly proud. Native birds occur here in abundance throughout the year, our rivers and lakes invite the migratory water fowls that annually pass this way, and the wooded streams and timbered hills offer shelter to countless birds that come here every year to nest and rear their young. The bird population is, of course, not as large as it formerly was, but while many valuable and beautiful birds have been reduced and driven to the verge of extinction, only a few species have as yet been entirely lost, and the general awakening of interest in bird life throughout the State marks the beginning of a new and brighter day for our feathered friends. We may confidently expect them in greater numbers as the forces for their protection are taken over by willing hands and carried into new and broader fields.

WILD TURKEYS.

Wild turkeys, like many of the larger native animals, formed no small part of the food supply of the pioneers, and in this way they have rendered a service of importance in the development of the State. These birds were at one time very numerous, but were so easily approached and shot that they were annually slaughtered in great numbers. It was not uncommon in the early days to see as many as five hundred of these splendid birds in a single bunch, and one hunter reports having seen a solid, unbroken acre of turkeys. Their habit of roosting in trees probably contributed in a large measure to their destruction. Hunters would locate a bunch, and after the turkeys had flown into the trees for the night, would walk quietly under them, where their large bodies outlined against the sky made an easy mark for a rifle. A man from Jefferson, with two companions, thus killed eighteen turkeys in one evening. A few wild turkeys are yet free in the mountainous regions of the State, but unless they receive legal protection it will be only a short time, at the present rate of destruction, until they are vanquished. Then we will have only the flock in the Wichita Preserve, which is as tame and fearless as poultry.

PRAIRIE CHICKENS.

We have left only a pitiful remnant of prairie chickens. They have been driven from the plains and rolling sand hills where they normally thrive, into wild retreats where they can secure some measure of protection from the constant illegal war that is being waged against them. A few flocks may yet be found in the Panhandle district, and a small number have in some way kept alive in Ellis County, but they represent only a small portion of what formerly existed here. Flocks of a thousand birds were sometimes reported, and a hunter with two companions once killed a wagon load of chickens on one hunt. The law provides no open season on these birds, and they may be brought back to an abundance where they can again be hunted, but it will take a long time and a more strict enforcement of the law than is now maintained.

QUAIL.

If the United States ever sees fit to erect a Hall of Fame for its birds, the quail will occupy one of the chief places, for it is by all odds the most valuable bird in North America. This is due not for its meat alone, although this is a delicacy that is seldom, if ever, surpassed, but for the constant, unselfish service which this bird renders to the farmer in its never-ending war against the ravages of insects and weeds. The quail is an ally of the farmer, staying at his post throughout the entire year and working on an average ten hours a day. In the summer it runs down and consumes as many as one hundred and fifty different kinds of insects, and in the winter it gathers and eats almost as many varieties of noxious weed seeds. This bird has diminished in number in Oklahoma, not at the hands of hunters alone, though they have contributed in a large measure, but also because farmers have thoughtlessly burned over its nesting and feeding grounds, and have made no provision for it in the new order on the farm. Every effort should be made to bring the quail back to its former abundance, and to anyone within the State who will provide suitable nesting places the State Game Warden will furnish free of charge a stock of these birds. There are a few counties where the quail is yet numerous enough to justify hunting it, but such places are exceedingly rare. Wherever this bird is diminishing, it is not only an economic blunder but a crime to decrease the number.

HAWKS AND OWLS.

Besides the quail we have a number of other birds that annually make a large contribution to the wealth of the State through the service they render to the farm and forest. Hawks and owls stand at the head of this list, but unfortunately the first of these two does occasionally take a chicken, and this causes it to be branded as a thief and an outlaw. No thought is given the excellent service which it renders through the months when it neither attacks nor cares to attack poultry. The average adult hawk consumes each day at least six mice or their equivalent in other rodents or harmful insects. Since we have learned by experiment that a single mouse causes an annual damage of at least two cents, it is a simple process to compute that each adult hawk is worth more than forty dollars a year to the farmers of the State. There are two exceptions—Cooper's hawk, and the sharp-shinned hawk. These species are thoroughly bad and deserve to be shot wherever found, but they look so much like beneficial hawks that it is unwise to hunt them in the field. A better plan is to have a gun at hand and be prepared to execute summary vengeance on any marauder that dares approach the chicken yard. Owls possess all the good habits of hawks, with none of the bad. They also have the additional advantage of working at night when the rodents are running about, so that it is difficult to compute the economic importance of the silent army of owls with representatives in every community within the State.

WOODPECKERS.

We have not less than ten different kinds of woodpeckers, and each of them represents a distinct value to the orchard and timber interests. They carry on an interminable search for the deadly grubs and other insects that burrow into and endanger the life of trees. Wherever a forest insect pest develops in any section it is sure to attract an army of hammering woodpeckers, and the birds are soon the victors. They also have played a part in spreading the forests. These birds are, in fact, real foresters, in that they are persistent planters of trees and do not give up their watchful care over them as long as they harbor a single insect pest. Yes, they do drill holes in telephone poles and riddle church steeples, but school boys carve their names on beautiful desks. Both are seeking to be perpetuated; there is no more reason for destroying one than the other.

EAGLES AND VULTURES.

People have never ceased to wonder at the matchless flight of eagles and vultures. They soar far into the upper atmosphere and float on motionless wings so smoothly that they appear as mere spots against the sky. It is said that these birds are never caught in the fury of the storms but fly above the clouds and descend only after the wind has spent itself. Both the golden and bald eagle frequent this State and have been known to nest in the mountainous regions of the south and east, but they occur in very small numbers and are only occasionally seen. The turkey and black vultures are found more abundantly and are commonly seen circling far above the earth in their constant search for carrion. They have keen powers of vision, and the characteristic "drop" of a vulture to a carcass will soon bring others from miles around.

WATER AND SHORE BIRDS.

Our State is situated within the migratory path of a very large percentage of all the water and shore birds that annually go north for the breeding and nesting season. Floating clouds of gulls are seen over the State during the spring and summer months, but on account of their powerful wings they do not often come to rest upon the ground. Solitary pelicans occasionally fly over, but they usually come in flocks. As many as one hundred and fifty of these feathered fishermen have spent the night on the Canadian River a few miles west of Norman. Ducks come and go throughout the entire year. On the protected lakes and ponds they become very tame and are often seen in great numbers. It is probable that more ducks occur in Comanche County than in any like area within the State. All told, at least thirty kinds of ducks frequent Oklahoma every year. Wild geese were once much more abundant than they are now, and while they do yet fly over the State in numbers, they seldom come down except in most inaccessible retreats, and it is a "red letter day" for the hunter who succeeds in bagging one of these birds. The towering form of the sandhill crane is sometimes seen along the larger streams, but its noble cousin, the whooping crane, has ceased to visit us. The great blue heron was never abundant here and its numbers now are very limited. However, we occasionally see this stilt-legged bird, with many of its small relatives, wading about the shallow ponds in search of crayfish and frogs. Its long beak makes it a dreaded hunter of the small streams, but Oklahoma will be the poorer if this ungainly but interesting bird ever disappears altogether from our shores.

CROWS.

Crows are a bad lot. They have a record of misdeeds almost as inky black as the coats they wear. Before the planting of grain fields gave them an unlimited food supply they were never numerous, but now they occur in numbers so large as to inflict heavy damage on the farmer in many sections of the State. Their fondness for sprouting corn is responsible for much of the trouble which they cause, but this is not all. Crows destroy large numbers of bird nests in the spring, and do not hesitate to feed the squawking fledglings to their own young. They come down in clouds on cane and kaffir fields and strip the heads before they can be gathered. This rapidly increasing multitude of destroyers presents a problem which will sooner or later have to be met. The crow has few natural enemies and is far too wise to be poisoned by the usual methods. A bounty is placed on their heads in many states, and this will probably soon be done in Oklahoma.

ENGLISH SPARROWS.

It is impossible to say anything too unkind about the English sparrow. There are few places in the State where this bird does not occur, and it is multiplying five times as rapidly as any other bird in North America. This bird is everywhere displacing the native birds, and is an invariable source of inconvenience about watering places and buildings. The English sparrow is not beautiful, cannot sing, and lives almost entirely on the hospitality of the farm. No one should lightly condemn a bird, but this one unquestionably deserves to die. These sparrows are gregarious in habit and can easily be poisoned in large numbers, so that any farm or city can entirely rid itself of these pests in one winter if it will only learn the methods and apply them.

SONG BIRDS.

Oklahoma is rich in the number and variety of her song birds. The larger part of these are, of course, migratory, and so remain here only a part of the year, but they make the woods merry with their songs from early spring until the horde of destructive insects is killed or driven to shelter by the approach of winter. These feathered songsters not only contribute immeasurably to the joy of living, which would be argument enough for their eternal protection, but they wage a ceaseless war against insect pests and weeds everywhere. They work at all hours when insects are about and pursue their prey into all accessible places. The robin is abundant here through the summer months, and is continually combing the yards and gardens in search of pests. It is a delight to watch this representative bird attempt to teach its half-grown young to shift for themselves, but they apparenly will not learn and after days of patient but vain effort the exasperated parent leaves part of a wriggling worm on the ground in front of the ravenous youngsters, and flies away. The young birds proceed in an awkward fashion to finish their meal, and with the awakening instincts are soon able to provide for themselves. During the winter months robins are so numerous in the southern part of the State that trees are often covered with the roosting birds, and vicious men go into these colonies at night with a lantern and when the frightened birds become bewildered and fly into the light they are struck down by the hundreds and their mangled bodies picked up and carried away for food. This condition applies, also, in a large measure to the cardinals. In the extreme northern part of the State this bird is migratory, but is a permanent resident of the central and southern part of Oklahoma, where it often gathers in great numbers. They feed very little on insect life, but subsist for the most part on weed seeds and wild fruits. Wherever they are found in large colonies they are preyed upon by degenerate and criminal men, who go about after night armed with lanterns, clubs, and sacks. If public sentiment in such communities is not strong enough to prevent this savagery, its discontinuance should be guaranteed by the State, even at the cost of a permanent patrol.

The present number of song birds can constantly be increased if the citizens will only exercise a little patience and tact. Birds are quick to recognize their friends and anyone who is willing to make the effort can double the number of birds on his premises in a single year. Many cities are organizing bird clubs and providing nesting sites and food for the birds that have suffered most on account of the change in nature occasioned by the coming of man. These communities will reap a reward in increased bird life that will repay the effort a thousand times. Aggressive effort for a worthy cause will ennoble the character of a girl or boy; it will create in the community a moral tone that is at once fine and beautiful. The first-hand study of wild life and its systematic care and protection offer possibilities to the public schools of the State that have not yet even been sounded.

One could not enumerate, much less discuss, in this brief account, all the song birds of Oklahoma. There are more than one hundred kinds, and every one is doing an excellent service for man and deserves every protection we can throw around it. Our bird life, fortunately, has not decreased so much as it has in some other states, and we may always have a wealth of feathered songsters if only we will it now.

FISHES AND MUSSELS.

FISHES.

Fishing in Oklahoma can never be successfully commercialized, for there are few large, permanent bodies of water within the State, and the depth and character of the streams are not such as to make possible the constant production of fish in quantities. Our rivers and creeks, however, contain a large number of excellent food fishes that will always furnish an abundance of wholesome sport for those who have the time and patience to go after them. Seining is legal in all the larger streams, and while too many fish are caught in this way, it probably will never materially lessen the supply, for the rivers are annually restocked from the larger streams during periods of high water. Some large fish are occasionally caught in Oklahoma. The yellow Mississippi cat is taken in many places, and specimens weighing as much as 100 pounds are not unknown. The blue and channel cat fish abound in nearly all our streams, but they are not so large. One weighing more than 50 pounds is extremely uncommon. Eels are found in most of the streams of the State. They range in size up to three feet in length, but the larger ones are rare. Among the scale fishes the buffalo, drums, and bass are the largest and most valuable, while the smaller ones including the goggle-eye, white, and sun perch are the most numerous and most widely distributed. The hickory shad is abundant in all the streams, but it is so thoroughly useless, and has choked so many uninitiated folk with its countless bones, that it does not deserve to be mentioned in an account of the fish resources of the State. Sturgeons and gars occur in all the larger streams and some attain considerable size. While we now have a fair amount of fish, they cannot always withstand the outrages being committed against them. Streams not more than 10 feet in width have been seined to the last fish, and one man was known to dynamite his own stream at spawning time! If, however, fishing is conducted in a sportsmanlike fashion, excellent and profitable sport can be assured and put on a permanent basis.

MUSSELS.

Fresh water clams or mussels occur abundantly in the Verdigris River throughout a large part of its course, and there is no reason why anyone living along this stream could not increase the resources of his farm by a little study and care in the propagation of these animals. Their shells are worth from $8 to $60 a ton, and the occasional finding of a valuable pearl adds an element of excitement and possibilities of fisherman has secured a pearl from that community which he had sold for $525. The mad excitement which this created did not fully spend itself until the last mussel from that region had been hauled from the water and examined. The clam industry probably never can be developed to large proportions in this State, but it can be made an additional source of revenue for those who have land located along suitable streams.

REPTILES.

SNAKES.

A fear once deeply grounded in the human mind can not easily be destroyed. This explains why we draw back with uncontrolled fright every time a tiny snake crosses our path. There are, of course, poisonous snakes, but we have only four of these in Oklahoma—the rattler, the copperhead, the cotton-mouth moccasin, and the coral snake. All our other snakes are no more poisonous than a rat or dog, and there is not nearly as much reason to fear them. The little spreading viper is totally devoid of any poison glands, and its peculiar habit of spreading its head is only a ruse to fool people, a ruse which to this day works well. The monster bull snake is as harmless as a pelican and worth a hundred times as much. It spends all its active life in search of gophers, rats, and mice, and is one of the most effective forces in holding these destroyers in check. This snake is the only foe of the pocket gopher, except the weasel, that can enter its burrow and pursue the occupant to certain death. It is not uncommon to find that a gopher caught in a trap has been half swallowed by one of these snakes, which found an impassable obstacle in the trap and gave up its victim. Two large bull snakes permitted to remain about a store-house will do more to keep down destructive rodents than five times as many cats. We have about 40 kinds of snakes in Oklahoma, but only the four named above are harmful. Poisonous snakes have distinctive characteristics than can be recognized at a glance. They usually have a head that is distinctly marked off from the body and does not taper gradually into the trunk region, as is the case with all harmless snakes. In addition, the rattlesnake, copperhead, and cotton-mouth moccasin have heavy depressions or pits between the eyes and the nostrils, and the tail is abruptly blunt. The rattler is a true sportsman. His bite is always deadly but he never strikes without first giving warning, and will eat no creature until its life is completely extinct. These snakes were at one time numerous over the prairies but they are now nearly extinct except in the rougher, unsettled regions. No one, of course, regrets the passing of this most deadly animal, but it is a pitiful condition that man insists on making an indiscriminate warfare on even the creeping things on the earth that do us no harm, and which must ever be sacrificed on account of a superstition and false fear that are centuries old.

TURTLES.

Besides the snakes, Oklahoma has a widely distributed reptile population that is both varied and interesting. The turtles and terrapins, of which we have more than 12 varieties, are cold-blooded creatures that lie dormant during the winter months, but are seen crawling or swimming everywhere in the summer. The soft-shelled turtle is especially abundant in many of the rivers, and the tortoises are familiar objects in the fields and gardens. The members of the turtle group in Oklahoma are for the most part rather small, but a few large ones are occasionally taken. Recently an alligator-snapping turtle weighing 110 pounds was taken from Blue River near Durant. When this specimen was discovered a member of the party ventured certain familiarities and extended the paddle end of an oar to within a few inches of its head. At once the massive jaws snapped away a portion of the oar and signaled a warning that was not to be disregarded. These larger reptiles are not abundant, and only a few are ever taken, for they frequent the deeper portions of streams and lakes, and are seldom seen except through periods of prolonged droughts when the retreating water leaves them stranded or in shallow pools.

LIZARDS.

The little blue-tailed lizards are present in large numbers and dart under some protecting object on the first approach of danger. Mountain-boomers are present in rough, hilly parts of the State where rocks abound, and the horny "toad," which is not a toad at all, but a lizard, is a common sight in practically every community of the State. A few small alligators have ascended the rivers as far as the central portion of Oklahoma. A specimen about four and a half feet long was taken from an over-flow lake south of Norman, and another was reported in a small lake near Durant. This is, of course, far outside the usual limits of their natural range, and it is one of the strange, interesting experiments of Nature in thus projecting an animal into a habitat remote from its established home, and under conditions that require radical changes in its manner of living if it is to survive.

INSECTS.

Why is an insect? This question has always been uppermost in the minds of farmers, while gardeners and fruit growers have never ceased to speculate on what a paradise this world would be if insects had never been created. Dwellers in antiquated hotels give eloquent testimony of the annoyance and humiliation caused by certain of these lowly animals, and the inhabitants of swampy regions know only too well of the deadly malaria that abounds and claims many human victims in regions that are infected with the Anophiles mosquito. Texas fever is transmitted through the bite of a tick, and is responsible each year for a loss in this State that cannot be expressed with less than six figures. The inconvenience and suffering resulting directly or indirectly from the housefly and the diseases which it spreads represent an economic loss that can scarcely be overestimated. Forests in all parts of the State are being weakened by the ravages of insect pests, and it is a common sight to see orchards destroyed and turned into cultivated fields, because the farmers could not compete sucessfully with the hordes of pests that find an easy food supply in the young fruit trees. Those people who were in the State in 1907 remember well how the greenbugs came early in the spring and settled like dust over the wheat fields. They remained until harvest time and destroyed about twenty percent of the wheat crop in Oklahoma. The destruction and waste occasioned by insects each year in this State amounts to more than $40,000,000. This is an economic waste that in time can be and must be controlled. Before the coming of the settlers there were no such things as insect plagues, but man has entirely upset the balance of Nature, and has made possible the unlimited production of these pests by killing or driving away their natural enemies, and providing in the cultivated crops a food supply that would not be possible in the wild order of things. Insects can be checked and reduced to proportions where they can inflict little harm. The United States Department of Agriculture has already provided and tested the methods, but one man working alone cannot do it; a county alone cannot, but the whole State working as an organized unit can, if it will, arouse itself, and by scientific methods strike forever from the ledger this annual waste of millions.

SUMMARY.

Communities in their regard for animal life undergo the usual cycle of development. There are at first the boyhood days, when the primitive instincts to kill and destroy are dominant, and indiscriminate slaughter is the order of the day. Buffaloes are shot from car windows, fish ponds are dynamited during spawning seasons, and song birds are sacrificed by the hundreds for rifle practice. We have now, happily, passed this barbarous age and are living in the young manhood period when some prudence is exercised and some restrictions imposed, but when the spirit of carelessness and indifference prevents the expression of some of the noblest impulses, and keeps us inactive even after we know the truth and have sensed our responsibilities. Let us hope that this State is soon to emerge into the period of mature manhood, characterized by a deep sense of our obligation to wild life, and a resolute determination to make reparation, as far as possible, for the blunders of the past, by preserving and increasing the wild things that yet remain, and to pass them on undiminished to the future generations.

CLASSIFIED LIST OF WILD ANIMALS IN OKLAHOMA.

The following is an annotated check list of the vertebrated animals occuring in Oklahoma. It is intended only as a preliminary list and is by no means complete. In its preparation the writer has been conservative and included only those species which, from authentic published records or personal experience, he knows are residents of this State.

Those animals which have recently been exterminated or occur only in captivity are indicated by an asterisk (*).

The order in which the species are listed is the one which was thought would be the most convenient for the average reader. The scientific names are given for the convenience of those who may wish to check this list with the animals in other states, where the common names differ widely from those of our own.

MAMMALS. CLASS MAMMALIA.

HOOFED ANIMALS. ORDER ARTIODACTYLA.

Buffalo. (American bison.)
Bison bison.
Whitetail deer. (Virginia deer.)
Odocoileus virginianus.
Antelope. (Prongbuck.)
Antilocapra americana.
*Elk. (Wapiti.)
Cervus canadensis.
Mule deer. (Black-tailed deer.)
Odocoileus heminous.

FLESH-EATING ANIMALS. ORDER CARNIVORA.

Mountain lion. (Puma. Cougar.)
Felis concolor.
Ocelot. (Leopard cat. Tiger cat.)
Felis pardalis.
Coyote. (Prairie wolf.)
Canis latrans.
Red fox.
Vulpes fulvus.
*Kit fox. (Swift fox.)
Vulpes velox.
Gray fox.
Urocyon cinereoargenteus.
Otter.
Lutra canadensis.
Weasel. (Long tailed weasel.)
Putorius longicauda.
Mink.
Lutreola vison.
Bobcat. (Wild cat. Red lynx.)
Lynx rufus.
Gray wolf. (Timber wolf.)
Canis occidentalis.
Common skunk.
Mephitis mesomelas.
Little spotted skunk. (Hydrophobia cat.)
Spilogale interrupta.
Badger.
Taxidea berlandieri.
Black Bear.
Ursus americanus.
*Grizzly bear. (Silver-tip grizzly.)
Ursus horribilius.
Raccoon.
Procyon lotor.

GNAWING ANIMALS. ORDER GLIRES. RODENTIA.

Beaver.
Castor texonsis.
Red squirrel. (Chickaree.)
Sciurus hudsonicus.
Gray squirrel.
Sciurus carolinensis.
Chipmunk. (Rock squirrel.)
Tamias striatus.
Striped ground squirrel. (Thirteen-lined, or Leopard Spermophile.)
Citellus tridecemlineatus.
Franklin ground squirrel. (Gray ground squirrel.)
Citellus franklini.
Prairie dog.
Cynomys ludovicianus.
Ground hog. (Woodchuck.)
Marmota monax.
Flying squirrel.
Sciuropterus querceti.
White-footed prairie mouse.
Peromyscus maniculatus bairdi.
Field mouse.
Microtus ludovicianus.
Harvest mouse.
Reithrodontomys dychei.
Small flying squirrel.
Sciuropterus volans.
Muskrate.
Fiber zibethicus.
Wood rat. (Pack rat.)
Neotoma macropus surberi.
Cotton rat. (Marsh rat.)
Sigmodon hispidus texianus.
Meadow vole.
Microtus austerus.
Wood vole.
Microtus pinetorum.
Pocket mouse.
Perognathus paradoxus.

Grasshopper mouse.
Onychomys leucogaster.
Kangaroo rat.
Microdipodops richardsoni.
Deer mouse.
Peromyscus atwateri.
Cotton mouse.
Reithrodontomys chrysotis.
Pocket gopher.
Geomys breviceps.
Prairie mouse.
Peromyscus canus.

RABBITS. FAMILY LEPORIDAE.

Cotton-tail.
Lepus sylvaticus.

Common jack rabbit. (Jack hare.)
Lepus texianus.

Whitetailed jack rabbit. (Prairie hare.)
Lepus campestris.

INSECT-EATING ANIMALS. ORDER INSECTIVORA.

Common mole.
Scadops aquaticus.

Prairie mole. (Silver mole.)
Scalops sp.
Short-tailed shrew.
Blarina brevicauda.

BATS. ORDER CHIROPTERA.

Little brown bat.
Myotis lucifugus.
Big brown bat.
Eptesicus fuscus.
Big-eared bat.
Myotis sublatus.
Tree bat. (Red bat.)
Lasiurus borealis.

POUCHED MAMMALS. ORDER MARSUPIALIA.

Virginia Opossum. ('Possum.)
Didelphis virginiana.

*Murine Opossum. (Mouse Opossum.)
Marmosa murina.

*A tropical species accidentally introduced in bananas. Has been taken at Yukon (2 specimens) and at Norman (13 specimens).

BIRDS. CLASS AVES.

GREBES.

Pied-bill grebe. ("Hell-diver." Dab-chick.)
Podilymbus podiceps.

LOONS.

Great northern diver. (Loon.)
Gavia imber.

GULLS AND TERNS.

Franklin's rosy gull.
Larus franklini.

Herring gull.
Larus argentatus.

Least tern.
Sterna antillarum.

CORMORANTS.

Double-crested cormorant.
Phalacrocorax auritus.

PELICANS.

Great white pelican.
Pelecanus erythrorphynchos.

DARTERS.

Snake-bird. (Darter. Water-turkey. Anhinga.)
Anhinga anhinga.

GEESE, DUCKS, AND SWANS.

American merganser.
Mergus americanus.

Hooded merganser.
Lophodytes cucullatus.

Mallard.
Anas platyrhynchos.

Black duck.
Anas rubripes.

Canvasback.
Marila valisineria.

Lesser scaup duck.
Marila affinis.

American golden-eye.
Clangula americana.

Buffhead.
Charitonetta albeola.

Gadwall.
Chaulelasmus streperus.
Baldpate.
Mareca americana.
Greenwinged teal.
Nettion carolinense.
Bluewinged teal.
Querquedula discors.
Shoveller.
Spatula clypeata.
Pintail. (Sprig.)
Dafila acuta.
Wood duck.
Aix sponsa.
Redhead.
Marila americana.
Greater scaup duck.
Marila marila.
Ringnecked duck.
Marila collaris.
Ruddy duck.
Erismatura jamaicensis.
Lesser snow goose.
Chen hyperboreus.
Blue goose.
Chen caerulescens.
Whitefronted goose.
Anser albifrons gambeli.
Canada goose.
Branta canadensis.
Hutchins's goose.
Branta canadensis hutchinsi.
Smaller whitecheeked goose.
Branta c. occidentalis.
Fulvous tree duck.
Dendrocygna fulva.
Whistling swan.
Olor columbianus.
Trumpheter swan.
Olor buccinator.

IBISES, HERONS, AND EGRETS.

Whitefaced glossy ibis.
Plegadis guarauna.
American bittern.
Botaurus lentiginosus.
Least bittern.
Ixobrychus exilis.
Great blue heron.
Ardea herodias.
Little green heron.
Butorides virescens.
*Snowy heron. (Snowy egret.)
Egretta candidissima.
Blackcrowned night heron.
Nycticorax nycticorax.
Yellowcrowned night heron.
Nyctanassa violacea.
Little blue heron.
Flo rida caerulea.

CRANES AND RAILS.

*Whooping crane.
Grus americana.
Little brown crane.
Grus canadensis.
Sandhill crane.
Grus mexicana.
King rail.
Rallus elegans.
Sora.
Porzana carolina.
American coot. (Mudhen.)
Fulica americana.
American avocet.
Recurvirostra americana.
Blacknecked stilt.
Himantopus mexicanus.

SNIPES.

Wilson's snipe.
Gallinago delicata.
Yellowlegs.
Totanus flavipes.

Pectoral sandpiper.
Pisobia maculata.
Least sandpiper.
Pisobia minutilla.
Semipalmated sandpiper.
Ereunetes pusillus.
Upland plover. (Bartramian sandpiper.)
Bartramia longicauda.
Marbled godwit.
Limosa fedoa.
Solitary sandpiper.
Helodromas solitarius.
Western willet.
Catoptrophorus semipalmatus inornatus.
Spotted sandpiper.
Actitis macularia.
Longbilled curlew.
Numenius americanus.
*Eskimo curlew.
Numenius borealis.
Wilson's phalarope.
Steganopus tricolor.

PLOVERS.

Blackbellied plover.
Squatarola squatarola.
Snowy plover.
Aegialitis nivosa.
Kildeer.
Oxyechus vociferus.

GROUSE, PARTRIDGE, ETC.

Bobwhite.
Colinus virginanus.
Western blue quail.
Callipepla squamata.
Prairie chicken.
Tympanuchus americanus.
Lesser prairie hen.
Tympanuchus pallidicinctus.

TURKEY.

Wild turkey.
Melleagris gallopavo.

PIGEONS.

Mourning dove.
Zenaidura macroura.
*Passenger pigeon.
Ectopistes migratorius.
Bandtailed pigeon.
Columba fasciata.

VULTURES.

Black vulture.
Catharista urubu.
Turkey vulture. (Turkey "Buzzard.")
Cathartes aura.

FALCONS AND HAWKS.

Mississippi kite.
Ictinia mississippiensis.
Cooper's hawk.
Accipiter cooperi.

Marsh hawk.
Circus hudsonius.
Sharp-shinned hawk.
Accipiter velox.
Red-tailed hawk.
Buteo borealis.
Harlan's hawk.
Buteo b. harlani.
Red-shouldered hawk.
Buteo lineatus.
Swainson's hawk.
Buteo swainsoni.

Goshawk.
Astur atricapillus.
Krider's hawk.
Buteo borealis kriderii.
Ferruginous rough-leg hawk.
Archibuteo ferrugineus.
Sparrow hawk.
Falco sparverius.
Osprey.
Pandion haliaetus.
Prairie falcon.
Falco mexicanus.

Rough-legged hawk.
Archibuteo lagopus.

EAGLES.

Golden eagle.
Aquila chrysaetos.

Bald eagle.
Haliaetus leucocephalus.

OWLS.

Barn owl. (Monkey-face owl.)
Aluco pratincola.
Short-eared owl.
Asio flammeus.
Barred owl.
Strix varia.

Screech owl.
Otus asio.
Burrowing owl.
Speotyto cunicularia.
Western horned owl.
Bubo virginianus pallescens.

PAROQUETS.

*Carolina paroquet.
Conuropsis carolinensis.

CUCKOOS.

Road-runner. (Chaparral Cock.)
Geococcyx californianus.

Black-billed cuckoo.
Coccyzus erythrophthalmus.

Yellow-billed cuckoo. (Rain-crow.)
Coccyzus americanus.

KINGFISHERS.

Belted kingfisher.
Ceryle alcyon.

Texas kingfisher.
Ceryle americana.

WOODPECKERS.

Hairy woodpecker.
Dryobates villosus.

Red-shafted flicker.
Colaptes cafer collaris.

Southern downy woodpecker.
Dryobates pubescens.
Red-bellied woodpecker.
Centurus carolinus.
Yellow-bellied sapsucker.
Sphyrapicus varius.
Red-headed woodpecker.
Melanerpes erythrocephalus.
Yellow-shafted flicker.
Colaptes auratus.
Pileated woodpecker.
Phloeotomus pileatus.
Texas woodpecker.
Dryobates scalaris.
Ivory-billed woodpecker.
Campephilus principalis.

GOATSUCKERS.

Whip-poor-will.
Antrostomus vociferus.
Chuck-will's-widow.
Antrostomus carolinensis.
Nighthawk.
Phalaenoptilus nuttallii.
Poor-will.
Chordeiles virginianus.

SWIFTS.

Chimney swift.
Chaetura pelagica.

HUMMING-BIRDS.

Ruby-throated humming-bird.
Archilochus colubris.

PERCHING BIRDS.

FLYCATCHERS.

Scissor-tailed flycatcher.
Muscivora forficata.
King-bird.
Tyrannus tyrannus.
Trail's flycatcher.
Empidonax trailli.
Olive-sided flycatcher.
Nuttallornis borealis.
Phoebe.
Sayornis phoebe.

Wood pewee.
Myiochanes virens.
Arkansas king-bird.
Tyrannus verticalis.
Crested flycatcher.
Myiarchus crinitus.
Yellow-bellied flycatcher.
Empidonax flaviventris.
Least flycatcher.
Empidonax minimus.
Green-crested flycatcher.
(Acadian flycatcher.)
Empidonax virescens.
Vermilion flycatcher.
Pyrocephalus rubinus.
Say's phoebe.
Sayornis saya.

LARKS.

Horned lark.
Otocoris alpestris.
Prairie horned lark.
Otocoris a. praticola.
Desert horned lark.
Otocoris a. lucolaema.

CROWS AND JAYS.

Blue jay
Cyanocitta cristata.
American magpie.
Pica pica hudsonica.
White-necked raven.
Corvus cryptoleucus.
American crow.
Corvus brachyrhynchos.

BLACKBIRDS AND ORIOLES.

Bobolink.
Dolichonyx oryzivorus.
Cowbird.
Molothrus ater.
Yellow-headed blackbird.
Xanthocephalus xanthocephalus.
Red-winged blackbird.
Agelaius phoeniceus.
Meadowlark.
Sturnella magna.
Western meadowlark.
Sturnella neglecta.
Orchard oriole.
Icterus spurius.
Baltimore oriole.
Icterus galbula.
Rusty blackbird.
Euphagus carolinus.
Brewer's blackbird.
Euphagus cyanocephalus.
Purple grackle.
Quiscalus quiscula.
Bronzed grackle.
Quiscalus quiscula aeneus.

FINCHES AND SPARROWS.

Purple finch.
Carpodacus purpureus.
American cross bill.
Loxia curvirostra.
American goldfinch.
Astragalinus tritis.
Arizona goldfinch.
Astragalinus psaltria.
Pine siskin.
Spinus pinus.
Lapland longspur
Calcarius lapponicus.
English sparrow. (House sparrow.)
Passer domesticus.
Vesper sparrow.
Pooecetes gramineus.
Western vesper sparrow.
Pooecetes gramineus confinis.
Savannah sparrow.
Passerculus sandvichensis savanna.
Western Savannah sparrow.
Passerculus s. alaudinus.
Cardinal.
Cardinalis cardinalis.
Rose-breasted grosbeak.
Zamelodia ludoviciana.
Western grasshopper sparrow.
Coturniculus savannarum.
Baird's sparrow.
Centronyx bairdii.
Lark sparrow.
Chondestes grammacus.
Harris's sparrow.
Zonotrichia querula.
White-crowned sparrow.
Zonotrichia leucophrys.
Chipping sparrow.
Spizella passerina.
Field sparrow.
Spizella pusilla.
Slate-colored junco.
Junco hyemalis.
Song sparrow.
Melospiza melodia.
Lincoln sparrow.
Melospiza lincolni.
Fox sparrow.
Passerella iliaca.
Towhee.
Pipilo erythrophthalmus.
Lasuli bunting.
Cyanospiza amoena.

Blue grosbeak.
Guiraca caerulea.
Western blue grosbeak.
Guiraca c. lazula.
Indigo bunting.
Cyanospiza cyanea.
Painted bunting.
Cyanospiza ciris.
Dickcissel.
Spiza americana.
Lark bunting.
Calamospiza melanocorys

TANAGERS.

Scarlet tanager.
Piranga erythromelas.
Summer tanager.
Piranga rubra.

SWALLOWS.

Purple martin.
Progne subis.
Cliff swallow.
Petrochelidon lunifrons.
Barn swallow.
Hirundo erythrogaster.
Bank swallow.
Riparia riparia.

WAXWING.

Cedar waxwing.
Bombycilla cedrorum.

SHRIKES.

Northern shrike.
Lanius borealis.
Loggerhead shrike.
Lanius ludovicianus.

VIREO.

Red-eyed vireo.
Vireosylva olivacea.
Warbling vireo.
Vireosylva gilva.
Yellow-throated vireo.
Lanivireo flavifrons.
White-eyed vireo.
Vireo noveboracensis.
Small white-eyed vireo.
Vireo n. micrus.
Black-capped vireo.
Vireo atricapillus.

WARBLERS.

Black and white warbler.
Mniotilta varia.
Worm-eating warbler.
Helmitheros vermivorus.
Golden-winged warbler.
Vermivora chrysoptera.
Nashville warbler.
Vermivora rubricapilla.
Tennessee warbler.
Vermivora peregrina.
Audubon's warbler.
Dendroica audubonii.
Pine warbler.
Dendroica vigorsii.
Parula warbler.
Compsothlypis americana.
Yellow warbler.
Dendroica aestiva.
Myrtle warbler.
Dendroica coronata.
Palm warbler.
Dendroica palmarum.
Ovenbird.
Seiurus aurocapillus.
Water thrush.
Seiurus noveboracensis.
Louisiana water thrush.
Seiurus motacilla.
Kentucky warbler.
Oporornis formosa.
Mourning warbler.
Oporornis philadelphia.
Northern yellow-throated warbler.
Geothlypis trichas.
Yellow-breasted chat.
Icteria virens.

Black-throated blue warbler.
Dendroica caerulescens.
Black-poll warbler.
Dendroica striata.
Black-throated green warbler.
Dendroica virens.
Cerulean warbler.
Dendroica caerulea.
Hooded warbler.
Wilsonia citrina.
Canadian warbler.
Wilsonia canadensis.
American red-start.
Setophaga ruticilla.
Wilson's warbler.
Wilsonia pusilla.
Magnolia warbler.
Dendroica magnolia.

PIPITS.

American pipit.
Anthus rubescens.
Sprague's pipit.
Anthus spraguei.

DIPPER.

American dipper.
Cinclus mexicanus.

THASHERS AND WRENS.

Mockingbird.
Mimus polyglottos.
Western mockingbird.
Mimus p. leucopterus.
Cat-bird.
Dumetella carolinensis.
Brown thrasher.
Toxostoma rufum.
Bewick wren.
Thryomanes bewicki.
House wren.
Troglodytes aedon.
Winter wren.
Nannus hiemalis.
Western marsh wren.
Telmatodytes palustris.
Rock wren.
Salpinctes obsoletus.

CREEPERS.

Brown creeper.
Certhia familiaris americana.

NUTHATCHES.

White-breased nuthatch.
Sitta carolinensis.
Slender-billed nuthatch.
Sitta c. aculeata.
Tufted titmouse.
Baeolophus bicolor.
Black-capped chickadee.
Penthestes atricapillus.

KINGLETS.

Golden crowned kinglet.
Regulus satrapa.
Blue-gray gnatcatcher.
Polioptila caerulea.
Ruby Crowned kinglet.
Regulus calendula.

THRUSHES.

Wood thrush.
Hylocichla mustelina.
Gray-cheeked thrush.
Hylociclia aliciae.

Wilson thrush.
Hylocichla fuscescens.
Hermit thrush.
Hylocichla guttata pallasii.
American robin.
Planesticus migratorius.
Bluebird.
Sialia sialis.
Western or mountain bluebird.
Sialia artica.

REPTILES. CLASS REPTILIA.

POISONOUS SNAKES.

Diamond-backed rattlesnake.
Crotalus adamanteus.
Prairie rattlesnake.
Crotalus confluentus.
Timber rattlesnake.
Crotalus horridus.
Massasauga.
Sistrurus catenatus.
Edward's massasauga.
Sistrurus catenatus edwardsii.
Cotton-mouthed moccasin.
Ancistrodon piscivorus.
Copperhead.
Ancistrodon contortrix.
Coral snake.
Elaps fulvius.

HARMLESS SNAKES.

King snake.
Ophibolus getulus.
Yellow-bellied king snake.
Ophibolus caligaster.
Corn or rat snake.
Coluber guttatus.
Pilot snake.
Coluber obsoletus.
Black snake. ("Blue racer.")
Zamenis constrictor.
Bull snake.
Pituophis sayi.
Coach whip snake.
Zamenis flagelliformis.
Blow snake.
Heterodon nasicus.
Green snake.
Cyclophis aestivus.
Common water snake.
Tropidonotus fasciatus.
Worm snake.
Carphophiops vermes.
Insect snake.
Diadophis amabilis.
"Dryland" moccasin.
Tropidonotus rhombifer.
Texas blind snake.
Glauconia dulcis.
Spreading adder.
Heterodon platyrhinus.

ALLIGATOR.

Alligator mississippiensis.

LIZARDS.

Blue-tailed lizard.
Eumeces quinquelineatus.
Sonora skink.
Eumeces obsoletus.
Clark's swift.
Sceloporus clarkii.
Horned lizard. (Horned "toad.")
Phrynosoma cornutum.

Common spotted lizard.
Holbrookia maculata.

Glass "snake."
Ophiosaurus ventralis.

Mountain boomer.
Crotaphytus collaris.

TURTLES AND TORTOISES.

Snapping turtle.
Chelydra serpentina.

Alligator snapping turtle.
Macrochelys lacertina.

Common musk turtle.
Aromochelys odoratus.

Keeled musk turtle.
Aromochelys carinatus.

Mud turtle.
Cinosternum pennsylvanicum.

Common box turtle.
Cistudo carolina.

Gopher tortoise.
Testudo polyphemus.

Agassiz's tortoise.
Testudo agassizi.

Southern soft-shelled turtle.
Trionyx ferox.

Spiny soft-shelled turtle.
Trionyx spinifer.

Diamond-back terrapin.
Malacoclemmys palustris.

AMPHIBIANS. CLASS AMPHIBIA.

TOADS.

Western spadefoot toad.
Scaphiolpus hammondii.

Common toad.
Bufo cognatus.

American toad.
Bufo americanus.

Toad.
Bufo compactilis.

FROGS.

Tree frog.
Hyla versicolor.

Cricket frog.
Acris gryllus.

Southern leopard frog.
Rana sphenocephala.

Southern tree frog.
Hyla squirrella.

Leopard frog.
Rana pipiens.

Green frog.
Rana clamitans

Common bull frog.
Rana catesbiana.

SALAMANDERS.

Tiger salamander. (Axolotl. Mudpuppy. Water-dog.)
Amblystoma tigrinum.

FISHES. CLASS PISCES.

Lake sturgeon.
Acipenser rubicundus.
Shovel-nosed sturgeon.
Scaphirhynchus platyrhynchus.
Mississippi blue cat.
Ictalurus furcatus.
Channel catfish.
Ictalurus punctatus.
Bullhead.
Ameiurus nebulosus.
Yellow cat.
Ameiurus natalis
Mud cat.
Leptops olivaris.
Common sucker.
Catostomus commersoni.
Hickory shad.
Dorocoma cepedianum.
Large-mouthed black bass.
Micropterus salmoides.
Yellow perch.
Perca flavescens.
Goggle-eye.
Ambloplites rupestris.
Sunfish.
Eupomotis gibbosus.
Long-nosed gar pike.
Lepidosteus osseus.
Black-gill sunfish.
Lepomis pallidus.
Buffalo.
Ictiobus cyprinella.
Common eel.
Anguilla chrisypa.

CHAPTER II.

PLANT LIFE IN OKLAHOMA.*

INTRODUCTION.

Oklahoma is an especially interesting region botanically, being a transition area between typical Mississippi Valley conditions in its eastern part, and very different conditions in the foot-hill region of the Rocky Mountains. The relation to the flora of the Gulf Plain and the more northern floras within its area is of much importance, since many species of both regions reach the limits of their ranges within the boundaries of the State.

Compared with the eastern states, Oklahoma is large, the area being 70,470 square miles, or more than the combined areas of the six New England states and New Jersey. Oklahoma extends from about 94° 30′ west longitude westward to 103° west longitude and from 33° 38′ north latitude to 37°. Its form is roughly rectangular with a narrow westward extension from the northwest corner, known as the Panhandle of the State.

PHYSIOGRAPHY AND GEOLOGY.

The lowest elevation in the State is in the southeastern corner, where the distance above sea level is about 400 feet. The surface rises gradually to the north and northwest, reaching an elevation of more than 4,000 feet in the west end of the Panhandle. Along the eastern edge of Leflore County, near the southeastern corner of the State, the high elevation of the hilly region, lying between the Ozark Mountains and the Ouchita Mountain uplift, extends from Arkansas into the State, the highest point being on Rich Mountain near Page, with an elevation of about 2,000 feet above sea level. Gradually lower elevations extend westward for about 50 miles. The northern third of the eastern part of the State lies in the western end of the Ozark range, and in some places along the State line the elevation reaches 1,000 to 1,500 feet.

About 30 miles north of Red River, and almost midway along the south side of the State, the Arbuckle Mountains are found, and extend for about 30 miles in a general east and west direction. In the southwestern corner of the State, approximately 30 miles from both the west line and Red River, the Wichita Mountains rise abruptly. These mountains consist of a series of hills and mountains of granite, porphyry,

*Stevens, G. W. and Shannon, C. W.—See "Introduction" page 5.

and limestone which extend southeastward nearly parallel with the general course of Red River for about 60 miles, gradually increasing in elevation to the east end of the range, where the greatest height is attained in Mt. Scott, some 1,500 feet in elevation. The other portions of the State have few local elevations except the conspicuous ranges of gypsum hills through several counties in the western part of the main body of the State (Pl. XI). These bluffs and butts often stand conspicuously above the surrounding regions, but their elevations do not exceed that of the more level country lying west of them.

West of the Gypsum Hills region, the surface is higher and more nearly even, except where cut into valleys and canyons by the nearly parallel rivers, and the surface of the region passes gradually into the Great Plains at about the 100th meridian.

Lying along the north side of the valleys of four rivers in the northwestern part of the State are belts of sand hills, ranging from 2 to 18 miles in width (Pl. VII). In the counties along the 100th meridian, the sandy regions, in some places, extend on both sides of the valleys, occasionally crossing the divides.

The surface of the strip 15 to 40 miles broad, extending from the southeastern corner of the State westward to about the middle of the south side, is a part of the Gulf Coastal Plain and consists of Cretaceous formations. The surface rocks of some areas in the Panhandle part of the State, and the western parts of Cimarron County, are also Cretaceous. The eastern half of the State, with the exception of the Ouachita and Arbuckle mountain uplifts, and the Cretaceous area along Red River, is included in the Carboniferous formations. The extreme northeastern part of the State, north of the Arkansas and east of Grand River, is composed of rocks chiefly of Mississippian age, while the remainder of the area designated consists of rocks of Pennsylvanian age. From about the middle line of the State, westward, with the exception of the Wichita Mountain region, and the High Plains area, the rocks belong to the Permian Redbeds. The surface rocks of the High Plains region consist principally of Cretaceous and Tertiary formations, with considerable areas covered by sands of recent age.

SALT PLAINS.

Low lying level sandy tracts varying from one acre to 50 square miles in area and having salt springs at or under the surface are known as salt plains. These are all in the western third of the State, and are mostly located far apart, although in three instances two or three are grouped in rather close proximity. The saline condition of these plains is due probably in all cases to the presence of salt springs.* However, the existence of such springs has not been demonstrated in the Great Salt Plains or United States Saline Reserve, in Alfalfa County; but the constant presence of salt water in the sandy floor even in the summer when evaporation is very great, strongly suggests the presence of saline springs

*Okla. Geol. Survey Bull. No. 11, Chap. VI. page 202.

beneath the floor of the plain. At this period the surface becomes white with an incrustation of salt. In most of the other salt plains the salt springs are flowing onto the surface, and the evaporation results in a crust of salt sometimes several inches thick on parts of the floor of the plain. One of these, in the valley of the Cimarron River, known as the Big Salt Plain of the Cimarron, beginning about 10 miles south of the Kansas line and extending about 8 miles down the river, is "the only place where rock salt can be obtained on the surface in all the plains country. This salt was known and used by the Indians, and was an article of trade from the Gulf to the British line, and this locality was a well-known geographical point from which distances were reckoned."* On several of the others salt is abundant and has been manufactured for local market at three or four of them.

The most conspicuous biological feature of the Salt Plains is the almost complete absence of life either in the water or soil. In some of the less concentrated ponds ditch grass (*Ruppia maritima*) is found. About their margins are found cat-tail (*Typha angustifolia*), bug-seed (*Corispermum hyssopifolium*), tumble-weed (*Cycloloma antriplicifolium*), sea-blite (*Suaeda linearis*), and water rush (*Scirpus campestris var. pludosus*); while sea purslane (*Sesuvium sesile*) grows in the salty sand of the floor plains. Salt grass, (*Distichlis spicata*), water-rush (*Ssirpus Torrey*), rush grass (*Sporobolus airoides*), and sand rache (*Atriplex argentea*) are found about the margins, but are occasionally likewise in other less alkaline parts of the State. Salt-marsh fleabane (*Pluchea camphorata*) is occasionally found in slightly alkaline situations about the plains and is not reported elsewhere in the State.

The size of the Great Salt Plain, about 12 miles long and 6 miles across in the broadest place, and its low position within a rather extended basin, lead to the belief that it is the site of an ancient lake which is called Lake Barde. The presence of the Least Tern nesting regularly in considerable numbers upon this plain and the Big and Little Salt plains of the Cimarron is taken by ornithologists as evidence of a former inland sea in the vicinity. This is a sea coast bird found rarely elsewhere except on the Florida, Gulf, and Lower California coasts.

GYPSUM HILLS.

Two formations of gypsum occur in the western part of the State (Pl. XI). These are exposed along the juncture of valleys and higher tracts, and mostly have the form of bluffs and buttes, the latter frequently called Glass Mountains or Gloss Mountains. The gypsum caps, on the bluffs or butts preserve the top of the elevation, with their slow decaying rock, while the underlying red clay is eroded more rapidly, often leaving perpendicular faces. (See Plate XX.) The rapid erosion of the sides of these hills and the excess of calcium sulphate (gypsum) on their tops make unfavorable conditions for plants. A few species are

*Okla. Geol. Survey Bull. No. 11, p. 205.

usually found on and about them and some of these are mostly not found elsewhere.

The most exclusively gypsum hill species are plains psilostrophe (*Psilostrophe villosa*), and phacelia (*Phacelia intergrifolia,*) and few-seeded mentzelia (*Mentzelia oligosperma*) which grow on the top of the gypsum cap or the last species in clefts in it; Nama (*Nama compactum*) seems to be confined almost exclusively to the guttered and weathered sides of these hills, the type locality of the species being the side of the butte in Major County known as Gloss Mountain. The sharp-scaled grass (*Erioneuron pilosum*) is common on somewhat more permanent places on the sides of the bluffs and is seen growing at the mouth of the cave in the illustration (Pl. XXI). Evening primrose (*Oenothera missouriensis*), sensitive brier (*Mimosa borealis*). aster (*Aster aricaefolius*), evening primrose (*Oenothera Hartwegi var. lavendulaefolia*), hairy purslane (*Portulaca polisa*), vetch (*Astragalus parviflorus*), and (*A. cuspidata*) are found about the sides and bases of the hills, but they also occur in other locations having red clay soil with a large content of calcium sulphate. The wild gooseberry (*Ribes aurea*), ill-scented sumac (*Rhus canadensis var. trilobata*), and red cedar (*Juniperus virginiana*) are often found rooting in the clefts of the gypsum cap, but are as often found in soil not especially rich in gypsum.

SAND-HILLS REGION.

Beginning about the middle of the State or a little farther westward there extends a strip of sandy country along the north side of each of four nearly parallel rivers, the Salt Fork of the Arkansas, Cimarron, North Canadian, and South Canadian (Pl. VII). These areas are from 2 to 18 miles broad and stretch northwestrly 50 to 210 miles. Near the western edge of the State the sandy area along the South Canadian broadens and becomes continuous on the north with that of Wolf Creek, a tributary of the North Canadian, and extending southward continues to Washita River. Farther southward along the western edge of the State extensive sandy regions are encountered on both sides of Salt Fork and North Fork of Red River, and east of the latter near its junction with Red River.

The sandy regions along the rivers in the northwestern part of the State seem to owe their origin to certain weather conditions which have maintained for a long period in the past, and to some extent are still operating.

These rivers are relatively long, the Cimarron and South Canadian having their sources in the Rocky Mountains. The rains in a semi-arid country are usually more rapid and of shorter duration than in better watered regions. In western Oklahoma formerly many rains lasted from 15 minutes to an hour, precipitating in that time from 1 to 4 inches of water. The uncultivated condition of the soil and the short duration of the rain resulted in most of the water running off and leaving, it is estimated, in some instances, as little as 10 per cent of it in the soil. The erosion caused by the resulting freshets and floods was great and the intensely muddy water of the streams carried much sand. The clay and dissolved substances were mostly carried on to larger rivers, but the sand was dropped in the river bed in the shape of bars as the velocity of the water diminished. Long periods of dry weather often intervened between rains, and in these drouth periods the rivers often became completely dry. During the summer dry periods the wind blows almost constantly from the south and has a velocity of from 10 to 40 miles per hour. The finer sand is carried by such wind and during most of the time, probably, since the Tertiary period, the winds have been strewing northward much of the sand brought down by the spring freshets. The area in the State thus covered by the sand from these four rivers is estimated to be between 3,000 and 4,000 square miles.

The average rainfall here has not increased in the last few decades, but the cultivation of the soil and growing of crops and trees since the settlement of the country have resulted apparently in the rainfall occuring more gently and for longer duration. This fact, together with the receptive condition of the soil caused by cultivation, results in a much larger part of the precipitation entering the soil and in fewer freshets and floods as well as less erosion. The sand supply of the river beds is no longer greatly increased and the sand is mostly kept moist during the summer and not carried by the wind. These rivers are now seldom dry and the areas of shifting sand are much reduced. In a few places the sand still drifts during the summer. In the southwestern part of Woods County, there are several square miles covered with shifting dunes and the region is advancing northward covering elm and cottonwood trees and causing Dog Creek to change its course.

The plant life of the sand-hills consists of fewer species than that of the adjacent non-sandy regions, the number being greatest along the margins of the sandy areas, in better watered places and on sand-hills where sand has long since stopped drifting. The Waynoka dunes, some of which are 75 feet high, have no plant life except an occasional top of a buried cottonwood still projecting above the drifting surface.

BLACK MESA REGION.

Where the Cimarron River enters the northwestern corner of the State it traverses a rather deep valley or broad canyon bounded by sandstone-capped bluffs and buttes on the south and by the Black Mesa on the north. The Black Mesa is a bed of lava or mal pais rock, now standing above the level of neighboring elevations. The flora is distinctly western.

SOILS.

The great variety of rock formations found in the regions above named produces soils differing much in character. The granites, limestones, sandstones, shales, and clays each make different types of soil.

The soils of the State may be grouped under the following heads: (1) Residual soils; (2) Transported soils. Under the second type are: (a) Colluvial soils; (b) Alluvial soils; (c) Aeolian soils. Residual soils are those which have been formed in the place where they are now found, and the soils of Oklahoma belong chiefly to this type. The transported soils of the State consist of those along the streams and the sand areas of the State. The eastern half of the State has limestone generally present in the soil. The western half has an abundance, and even an excess of gypsum, even in the sand hill areas which are in the general gypsum region. The ground water also contains gypsum, and wells in the Permian red clay often furnish water so strong with gypsum that farmers must depend upon cisterns for holding water for domestic purposes. A large percentage of the soils of the State are alkaline. The percentage of acid soils is relatively small.

PRECIPITATION.

The rainfall is greatest in the southeastern part of the State, where it averages about 43 inches per year, gradually diminishing westward to Cimarron County, where the average is only 15 inches per year. A considerable part of the central-western part of the State also has a scanty rainfall. For further information concerning the precipitation, the reader is referred to a discussion of rainfall on page 37.

Lines marking places of equal rainfall extend mostly a little west of south from the northern boundary of the State.

LIFE ZONES.

Four life zones have been distinguished in Oklahoma. The Upper Austral covers a small area at the middle of the north edge and most of the Panhandle. The Carolinian zone extends over the rough and elevated regions in the eastern part of the State. The Lower Austral zone covers the central and a part of the eastern portions, while about two-thirds of the eastern portion is in the Austroriparian. These regions pass imperceptibly into each other, except in a few cases, where the change is rather abrupt (the Carolinian having rather distinct lines of contact with the Austroriparian).

RELATION OF PHYSICAL CONDITIONS TO THE FLORA.

In Oklahoma the problem of determining the relation of the different types of soil, of the elevation (life zones), of rainfall, and of distributing agents to the distribution of our species is complicated and difficult. By reference to the maps contained in this report, it will be seen that the lines showing the limits of the geological formations, and lines of equal rainfall largely coincide, and that the lines of equal elevation also in a general way coincide with these. Along the line of junction between the Pennsylvanian and Permian formations, numerous species reach the limit of their western range in the State. Conspicuous among these are Sycamore (*Platanus occidentales*), hickory (*Corya Illinoensis*), hickory (*Carya cordiformis*,) hackberry (*Celtis mississippiensis*), blackberry (*Rubus villosus*), black-cap raspberry (*Rubus occidentalis*), dwarf chestnut oak (*Quercus prinoides*), black oak (*Quercus veluntina*), southern red oak (*Quercus texana*), blue ash (*Fraxinus quadrangulata*), persimmon (*Diospryros virginiana*), winged elm (*Ulmus alata*), flowery dog wood (*Cornus florida*). Several of these extend somewhat father westward than the Pennsylvanian-Permian line in the southern part of the State. Since the lines of equal precipitation swing westward toward the southern end of the State, and the species named are found in their farthest west stations in places suffering less from evaporation than the average conditions of the regions (mountains, canyons, and river valleys), it seems that the governing factor of their western limits is the amount of moisture accessible in the soil and the relative freedom from excessive evaporation, rather than the nature of the soil.

There is a gradual thinning out of eastern species toward the western part of the State, the western limits of their ranges often extending as slender tongues along the valleys of the larger rivers, a fact which seems to point to the reduced moisture supply as the determining factor limiting their ranges. Prairie crab apple (*Pyrus ioensis*), Amercian hazelnut (*Corylus americana*), beech (*Fagus grandifolia*), hop hornbeam (*Ostrya virginiana*), cucumber tree (*Magnolia acuminata*), smooth alder (*alnus rugosa*), chinquapin (*Castanea pumila*), willow oak (*Quercus phellos*), water oak (*Quercus nigra*), are found but the division of the Carboniferous formations where they grow terminates mostly in a zone of precipitation between 37 and 40 inches, and since the range of most of these species includes regions of unlike soil conditions, but not any territory with less than 37 inches of annual rainfall, it seems to follow that the amount of precipitation is the governing condition. Aside from certain plans characteristic of special conditions to be described later, the delimiting factor in most of our plant ranges seems to be the amount of rainfall.

RELATION OF OUR FLORA TO THAT OF ADJACENT REGIONS.

The examination of the ranges of 1,000 of our species chosen as representatives shows the following facts concerning the relation of our flora to that of adjacent regions. This number of species was chosen rather than the entire number (about 1500) because the ranges of about one-third of them are broken, very irregular, or of such outline, such as introduced and vestigial species, as not to lend evidence in the present investigation. The results are given in the number of species per hundred of our range, minor fractions being omitted, and major fractions used as integers.

The flora is divided into groups as follows: (1) *Northern flora*, consisting of species whose southern range does not extend farther southward than Oklahoma (or rarely slightly farther), and divided into two groups: (a) species ranging through Upper Austral or the Carolinian and into the Transition or Alleghanian; and (*b*) species whose northern range limits lie in the Upper Austral or Carolinian. (2) *Mid-Mississippi Valley flora,* species ranging well into both the Upper Austral or Carolinian and the Lower Austral or Austroriparian. (3) *Southern flora,* consisting of species whose ranges are confined to the Lower Austral and Austroriparian, and divided into two groups: (a) those species ranging mostly through the lower Mississippi Valley, and (b), those species having ranges confined to Oklahoma, eastern Texas, and Louisiana. (4) *Western species* placed into two groups: (a) those whose ranges extend westward beyond the Upper Austral and into various zones beyond, and (b) those species confined to the Upper Austral zone (Great Plains). (5) Local species: (a) those species whose ranges do not extend beyond any adjacent state, and (b) those whose ranges are confined to Oklahoma. There is some duplication of species in division (3) group (b), and division (5) group (a), but the number is less than one per cent.

Approximately 5 per cent of our species are found throughout the State, or in other words, 19 out of 20 of our plants reach the limit of their ranges in Oklahoma. 70 per cent of our flora is the western extension of Mississippi Valley flora, and 7 per cent more is doubtless derived from the flora of that region. Twenty-one per cent is derived from the Great Plains or more westerly regions, and there are reasons for believing that about 2 per cent probably had local origin in Kansas, Oklahoma and Texas.

DISTRIBUTING AGENCIES.

The importance of distributing agencies is exaggerated in Oklahoma by the fact that 95 per cent of our species do not extend entirely over the State. The present distribution of many species is comparatively recent, for the details of their ranges are determined by local soil and water conditions that have been developed within a comparatively brief past. The western limits of many of our eastern species have numerous slender extensions or outlying stations in the regions west of them, these being mostly strips following the valleys and canyons of streams up their courses.

Two of the most generally distributed species in the State are Carolina poplar (*populus deltoides*), and long-leafed willow (*Salix longifolia*). Their wide range depends upon the facts that appropriate soil and moisture conditions for them exist over most of the State, and that the species have been able to reach the stations where now found. The wind is doubtless the agency responsible for the wide distribution of these, as well as for numerous others blown as tumble-weeds or whose seeds have pappus or coma of such nature as to enable them to be carried by the wind, notable examples being found among our composites and milkweeds.

As far as our eastern species are concerned, the consideration of water as a distributing agency may be discarded. Our rivers all flow south-easterly, and while they doubtless contribute to the distribution of some of our western species, they cannot assist those migrating westward.

Oklahoma is far enough south (33° 38′ to 37°) so that many birds which are only summer residents farther northward, such as Robins, Bluebirds, and Waxwings are common here in the winter, and subsist largely or partly upon various kinds of dried fruits. In summer many species of berry-eating and drupe-eating birds assist in the distribution, and as a result one or more species of hackberry (*Celtis*), plum (*Prunis*), grape (*Vitis*), Virginia creeper (*Parthenocissus*), soapberry (*Sapindus*), sumac (*Rhus*), greenbrier (*Smilax*), pokeweed (*Phytolacca*), Cedar (*Juniperus*), oak (*Quercus*), prickly pear (*Opuntia*), and *cactus* (*Mammillaria*) have found their way up valleys, into tributary and adjacent valleys or far out on the prairies.

The examination of numerous stomachs of cedar waxwings, robins, bluebirds, blackbirds, tanagers, orioles, mockingbirds, brown thrashers, catbirds, and thrushes reveals seeds and nuts of fruits that help make up the birds diet. The red cedar high on the canyon bank, the soapberry tree at the head of a tributary canyon or on a hillside above other trees, the solitary grape vine or sand plum in a grassy valley far removed from others of its kind, the skunk-bush (*Rhus canadensis var. trilobata*) perched on the brow of a gypsum-capped butte, the Missouri currant among the rocks on the rough side of a glass mountain, the lone hackberry in a prairie draw, the rock-grape growing in a cleft in a sandstone or granite ledge, are monuments to the services of birds as distributors of plants.

The writer has been much interested in the relation of the wild turkey to the distribution of the black-jack (*Quercus marilandica*) *and* Texas oak (*Q. undulata*). In years when these oaks bear fruit, their acorns have supplied a large proportion of the food of the wild turkeys in the vicinity, and the forests of the former species (black-jack oak) have been most favorable places for the hunter of these birds in winter. Examination of their crops in winter reveals surprisingly large numbers of acorns, often more than a hundred being found in a single bird.

The turkey is a ground-living bird and often travels in flocks for several miles on foot, passing frequently across divides between valleys. Until 20 or 25 years ago this bird was abundant in the semi-timbered and prairie regions of Oklahoma and Texas. The accidental death of one of them often resulted in leaving dozens or scores of acorns on the ground miles from oak trees. Some of these acorns must have been in condition for germination and grown into trees, and the oaks' range was thus extended. In the sandhill region along the north side of Cimarron River in Woods County the black-jack oak reaches its northwestern limit in the State. A similar region along the North Canadian has this tree reaching its last station 30 or 40 miles farther southwestward. Along the borders of the wooded regions in these sections and especially just at the western edge and often from a half to two miles beyond one finds places where numerous small oaks grow very closely together over a few square rods of area showing by their cross-sections about the same age. One finds no such clusters of quite young trees, nor does one find the large oaks in the woods conspicuously clustered. The explanation seems to be that the cluster of young oaks is from acorns left where a coyote, wild-cat, or eagle destroyed a turkey. Most of the trees in the clusters that have been formed for centuries have disappeared in the struggle for existence, and the cluster as such is no longer evident. There are no new outlying clusters now appearing because, it is believed, there are no longer wild turkeys in these regions.

PLANT FAMILIES IN OKLAHOMA.

The following pages give a summarized discussion of the plant life of Oklahoma. The summary is by families, and under each family name, the more common representatives are given, also general statements in regard to habitat, character of growth, distribution, and importance. In some cases where several species are known, the number is indicated by the figure (7) within the parenthesis directly following the name.

FERNS AND AILLIES (PTERIDOPHYTA.)

FERN FAMILY—(*Polypodiaceae*).—Including a large number of ferns, among which are the scaly, beach, powdery, maidenhair, bracken and lip, lady, walking, christmas, shield, woods, evergreen, wood bladder, and sensitive ferns, also several varieties of the royal fern family, adder's tongues, water ferns, the horsetail or scouring rush family, and quillworts.

SEED-BEARING PLANTS—FLOWERING PLANTS (SPERMATOPHYTA).

GYMNOSPERMAE.

PINE FAMILY—(*Pinaceae*).—Includes the pines, cedars (3), juniper, and water cypresses (in swamps in southeastern corner of the State).

ANGIOSPERMAE.

CATTAIL FAMILY—(*Typhaceae*).—The broad-leaved cattail is found in wet places throughout the State, while the narrow leaved or salt plains cattail is found only in wet places in the northwestern part of the State about the salt plains.

BUR-REED FAMILY—(*Sparganiaceae*).—Includes the bur-reeds found in moist or wet places in the eastern half of the State.

POND-WEED FAMILY—(*Naiadoceae*).—Includes the pond-weeds and ditch grasses, water-plantin family (*Alis maceae*). This family includes the arrowheads (*Sagittaria*) (9), the burr-heads (*Echinodorus*), and water-plantins (*Alisma*).

GRASS FAMILY—(*Gramineae.*)—This family includes a very large number of grasses, both wild and cultivated, including the Johnson grass, foxtail, reed grass, blue stems, wild oats, and many others.

SEDGE FAMILY—(*Cyperaceae*).—Includes the cypresses, found in wet sandy soils over most of the State, nut grasses, spike rushes (36), bullrushes (9), umbrella grasses, dwarf sedges, beak-rushes, nut-rushes, and sedges, of which there are a large number.

ARUM FAMILY—(*Araceae*).—This family includes the Indian turnip, or Jack-in-the-pulpit, found in moist woods in the eastern part of the State. The green dragon or dragon root is found in moist woods over the State, except the west and northwest part; and the sweet flag or calamus-root is found in the eastern part of the State.

DUCK WEED FAMILY—(*Lemnaceae*).—This family includes several species of small green plants, floating in large numbers on quiet water, found to some extent over the State, but chiefly in the eastern half.

YELLOW-EYED GRASS FAMILY—(*Xyridaceae*).—Includes only one genus found in the State. This is the slender yellow-eyed grass in the eastern part of the State.

SPIDERWORT FAMILY—(*Commelinaceae*).—Herbs with jointed round stems, mostly branched, including the spiderworts and day-flowers.

PICKEREL WEED FAMILY—(*Pontederiaceae*).—Including the picarelle, wood and mud plantins. The plants of this family grow in the central-eastern part of the State.

RUSH FAMILY—(*Juncaceae*).—Includes the rushes (20).

LILY FAMILY—(*Liliaceae*).—This family includes the bunch flower, false helebore, wild onions (including the garlics), and Oklahoma lily—found on prairies over the entire State, except part of northeast corner, lilies, dog-tooth violet, bellworts, hyacinths, star flower, yuccas (soap root), asparagus, false Solomon's Seal, Solomon's Seal, lily of the valley, wake robin (*Trillium*), and Green briars (7).

YAM FAMILY—(*Dioscoreaceae*).—Wild yam root (Colic root).

AMARYLLIS FAMILY—(*Amaryllidaceae*).—Yellow star-grasses.

IRIS FAMILY—(*Iridaceae*).—Irises, blue flags, and blue-eyed grasses.

ORCHARD FAMILY—(*Orchidaceae*).—This family includes the mocassin flower, the lady's slippers and showy orchis, and several other species.

LIZARD-TAIL FAMILY—(*Saururaceae*).—Lizard-tail in swamps and along pond margins in eastern part of State.

WILLOW FAMILY—(*Salicaceae*).—Including the willows of which there are at least 6 species and the poplars, including the silver-leaved or white poplar and the cottonwoods (2).

WALNUT FAMILY—(*Juglandaceae*).—The walnuts (3) and hickories (10).

BIRCH FAMILY—(*Betulaceae*).—Including the hazel, iron wood, hornbean, birches, and alders.

OAK FAMILY—(*Cupuliferae*).—Including the beech, chestnuts (2), and oaks (17).

NETTLE FAMILY—(*Urticaceae*).—Including the elms (3), hackberries (2), Bois d'Arc (Osage orange), mulberries, woodnettles, stinging nettles, and hops.

SANDAL-WOOD FAMILY—(*Santalaceae*).—Mistletoe. The toad-flaxes, partly as root parasites on trees and shrubs.

DUTCHMAN'S-PIPE FAMILY—(*Aristolochia macrophylla*).—

JOINT WEED FAMILY OR BUCKWHEAT FAMILY—(*Polygonaceae*).—Includes the gray-weeds, docks and sorrels, and joint weeds.

GOOSE FOOT FAMILY—(*Chenopodiaceae*).—Including the sand tumble weed, small cypress, lamb's quarters, mealy goosefoot, oak-leaved goosefoot, dwarf lamb's quarters, Russian thistle, Russian lamb's quarters, and several others; white sage (winter-foot), bug-seed, Jerusalem oak, Mexican tea.

AMARANTH FAMILY—(*Amaranthaceae*).—Includes Amaranths pig-weeds, tumble weeds, western water hemp, and blood-leaf (*Juda's* bush).

POKE-WEED FAMILY — (*Phytolacaceae*). — The poke-weeds (poker-root), common in wet and moist places throughout the State, except the Panhandle counties.

FOUR O'CLOCK FAMILY—(*Nyctaginaceae*).—Including several species of the four o'clock, generally widely distributed.

CARPET WEED FAMILY—(*Aizoaceae*).—Western sea purslane and carpet weeds.

PINK FAMILY—(*Caryophyllaceae*).—The pearl worts, sand worts, chickweeds, corn cockle (corn rose), catchfly, fire pink, starry campion, and bouncing bet.

PURSLANE FAMILY—(*Portulacaceae*).—Spring beauties, common throughout the State, fairy flower, talinums, purslanes, and rose moss.

HORNWORT FAMILY—(*Ceratophyllaceae*).—Hornweeds or hornworts, common in ponds and slow streams in the eastern part of the State.

WATER LILY FAMILY—(*Nymphaeaceae*).—Includes the cow lily (spatterdock), water lilies, American lotus (Chinquapin, water yuankapin), and water target.

CROWFOOT FAMILY—(*Ranunculaceae*).—Including the butter cups (crows-foot), meadow rues, anemone, wild clematis, leather flowers, columbine (rock-bells), lark-spurs, and baneberry.

MAGNOLIA FAMILY—(*Magnoliaceae*).—Magnolia, includes several species, only one of which is native to the State. This is the *magnolia accuminata* (Cucumber tree).

CUSTARD APPLE FAMILY—(*Anonaceae*).—Pawpaws, (*Asimina triloba*) in wooded areas in the northeastern corner of the State.

MOON SEED FAMILY—(*Menispermaceae*).—Includes the moonseeds and cupseeds. These plants grow along streams, and are widely distributed, except are not found in the western part of the State.

BARBERRY FAMILY—(*Berberidaceae*).—This family includes the may-apple or mandrake. This is found in rich woods in the eastern part of the State.

LAUREL FAMILY—(*Lauraceae*).—Includes the sassafras (ague-tree) and spiced bush. The plants of this family are limited to the eastern part of the State.

POPPY FAMILY—(*Papaveraceae*).—Includes the blood-root, found in rich woods in the eastern part of the State, and the prickly poppies on prairies and sandy soils, widely distributed over the State.

FUMITORY FAMILY—(*Fumariaceae*).—Includes the corydalis, of which there are several varities found chiefly in the eastern half of the State.

MUSTARD FAMILY—(*Cruciferae*).—Includes the whitlow grasses, velvet-flowers, bladder-pods, pepper-grasses (including the tongue grass, penny-cress), shepherd's purse, false flax, horseradish, mustards, turnip, yellow phlox, water-cresses, purple rocket, bitter-cresses, and rock-cresses.

CAPER FAMILY—(*Capparidaceae*).—Clammy-weeds and spider flower.

ORPINE FAMILY—(*Crassulaceae*).—Includes the stonecrop, live-for-ever and rock-moss.

SAXIFRAGE FAMILY—(*Saxifragaceae*).—Texas saxifrage, alum roots, mock-orange, wild hydrangea, Virgina willow, Missouri currant (buffalo-currant).

WITCH-HAZEL FAMILY—(*Hamamelidaceae*).—The witch-hazels, sweet gum (star leaved gum, red gum, bilsted), in woods in eastern and southeastern part of the State, sycamore (buttonball tree, buttonwood plane tree), along streams. This last known tree is found along the streams in the eastern half of the State, and westward to the Chickaskia River, along the north side.

ROSE FAMILY—(*Rosaceae*).—American ipecac, in the extreme eastern edge of the State; apple trees including the western crabapple, on river bottoms in the northeastern part of the State; service berry (shad-bush) over the eastern edge of the State, westward to the Arkansas River in the northern part.

HAWS OR THORN FAMILY—(*Crataegus*).—Thorns (11). Wild strawberries, cinquefoils, avens, small-leaved nut, mahogany (on sandstone buttes in Cimarron County), blackberries, raspberries, dewberries, agrominy, wild roses, (6), wild plums, wild berries (10).

PEA FAMILY—(*Legiminosae*).—This family includes the sensitive briar found on rough hill sides and grassy slopes throughout the State, except in the extreme western part; Desmanthus (3), mesquite (on prairies and river valleys in the western third of the State); Kentucky coffee tree (coffee-bean tree), honey locust, blueweed or cycle pod, partridge pea, sensitive plants, bank burr, redbud, Judas tree, wild indigo, and false indigo, rattle boxes, Nebraska bluepine (western Panhandle counties), clovers (trifolium), sweet clovers, prairie bird's foot, trefoil, snakeroot, Indian or prairie turnip, also called bread-root, bead plant, or devil's shoestring, river locust or false indigo, wild sweet pea, western indigo plant, peatree, wisteria, ground plums, tick-trefoil (15), bush-clovers (12), pencil flower, vetches (8), wild peas, ground-nut, wild bean, hog peanut, milk peas.

FLAX FAMILY—(*Linaceae*).—Includes the cultivated and wild flaxes (8).

OXALIS FAMILY—(*Oxalidaceae*).—This family includes the oxalis or wood sorrels. All of the forms are found throughout the State, but are most abundant in the central and eastern part.

GERANIUM FAMILY—(*Geraniaceae*).—Crane's bill geraniums, in waste places and lawns, over most of the State, but chiefly in the eastern part.

CALTROP FAMILY—(*Zygophyllaceae*).—Herbs having prostrate, much branched stems, including the coltrap burr, or Mexican sand burr, found in the northwestern part of the State; and the greater caltrop found throughout the prairie parts of the State.

RUE FAMILY—(*Rutaceae*).—Includes the prickly ash or toothache tree, found in the eastern and southern parts of the State; and the hop tree (*ptelea trifoliata*), found in open woods and rocky banks throughout the State.

MILKWORT FAMILY—(*Polygalaceae*).—Includes 5 species of the milkwort, widely distributed throughout the State.

SPURGE FAMILY—(*Euphorbiaceae*).—This family includes several plants, among which are the spurge nettles (tread-softly), the croton weeds (goat weed), the three-seated mercury, queen's delight, and the spurges (23).

SUMAC FAMILY—(*Anacardriaceae*).—Includes the sumacs, poison ivy and American smoke-tree (eastern part of the State).

HOLLY FAMILY—(*Ilicaceae*).—Includes the hollys of which 2 species are found in Oklahoma, the American white or Christmas holly, and the swamp holly, found in swamps and thickets in the eastern part of the State.

STAFF TREE FAMILY—(*Celastraceae*).—Includes the wahoos (burning bush) and bittersweets.

BLADDERNUT FAMILY—(*Staphyleaceae*).—The true bladdernut is represented by only one species in our State, the American bladdernut found in woods and thickets in the northeast corner of the State.

MAPLE FAMILY—(*Aceraceae*).—Several species found growing chiefly in the eastern part of the State, also many varieties in cultivation. (See list of trees and shrubs following).

SOAPBERRY FAMILY—(*Sapindaceae*).—This family is represented by the chinaberry or soap berry tree, and three or four species of the buckeye.

FOR-GET-ME-NOT FAMILY—(*Balsaminaceae*).—This family is represented only by the spotted or wild touch-me-not, sometimes called silver leaf, found in moist places in the eastern part of the State.

BUCKTHORN FAMILY—(*Rhamnaceae*).—This family includes chapparel (condilia), supple jack (rattan vine), Carolina buckhorn, and New Jersey Tea or redroot.

GRAPE FAMILY—(*Vitaceae*).—This family includes the Virginia creeper, which is common throughout the State; the false grapes, pepper vine, and several species of wild grapes.

LINDEN FAMILY—(*Tiliaceae*).—This family includes the basswood or linden tree of which there is but one species growing native in the State.

MALLOW FAMILY—(*Malvaceae*).—Includes the Indian mallow, false mallow, running mallow, poppy mallows (6), and rose mallows (3).

SAINT JOHN'S-WORT FAMILY—(*Hypericaceae*).—This family includes St. Andrews' cross and the St. John's-wort of which there are several species, found chiefly in the eastern part of the State.

ROCK-ROSE FAMILY—(*Cistaceae*).—This family includes the frost weeds and the pin weeds, found growing in dry, sandy places in the eastern two-thirds of the State, and also in the vicinty of the Wichita and Arbuckle mountains.

VIOLET FAMILY—(*Violaceae*).—There are about 20 species of violets known in the State. Some of these are widely distributed, but are chiefly found in the eastern part of the State.

PASSION FLOWER FAMILY—(*Passifloraceae*).—This family is represented by 2 species found in the eastern part of the State.

LOASA FAMILY—(*Loasaceae*).—Consisting of small, much branched herbs with rough surfaces and with barbed stinging hairs, found on sandy hill sides and prairies throughout the State.

CACTUS FAMILY—(*Cactaceae*).—This family is represented by the hedgehog, pin cushion, cacti, and the prickly pears.

LOOSESTRIFE FAMILY—(*Lythraceae*).—This family includes the water purslane, tooth cups, loosestrifes and wax-weeds.

MEADOW BEAUTY FAMILY—(*Melastomaceae*).—Includes the meadow beauties found growing in sandy, swampy places in the eastern part of the State.

EVENING PRIMROSE FAMILY—(*Onagraceae*).—Includes various willow-herbs, the evening primroses (19), gaura, and enchanters night shade.

WATER MILLFOIL FAMILY—(*Haloragidaceae*).—These are aquatic herbs, including the water millfoil and the mermaid weeds.

GINSENG FAMILY—(*Araliaceae*).—This family is represented in our State only by the common ginseng (*panax quinquefolia*), found in rich woods in the eastern part of the State.

PARSNIP FAMILY—(*Umbelliferae*).—This family is represented in our State by several species of the snakeroot, pennyworts, chervil, sweet cicely, sand parsley, cowbane, hedge parsley.

DOGWOOD FAMILY—(*Cornaceae*).—The dogwoods, small trees or shrubs, represented in this State by 4 species and the tupelos, represented by tupelo gum, and the sour or black gum.

LOGANIA FAMILY—(*Loganiaceae*).—A family represented by only 2 species in our State, so far as known.

GENTIAN FAMILY—(*Gentianaceae*).—This family includes the rose-pinks and centaurys.

DOGBANE FAMILY—(*Apocynaceae*).—Includes amsonia, dogbane and Indian hemp.

MILKWEED FAMILY—(*Asclepiadaceae*).—This family includes the milkweeds, of which there are at least 20 species, the sand vine, and baldwin's anglepods.

MORNING-GLORY FAMILY—(*Convolvulaceae*).—This family includes the morning-glories, of which there are 10 or 12 species, widely distributed over the State; the bindweeds and dodders.

PHLOX FAMILY—(*Polemoniaceae*).—The family includes the phloxes and gilias.

WATERLEAF FAMILY—(*Hydrophyllaceae*).—Includes phacelia and nama.

BORAGE FAMILY—(*Boraginaceae*).—This family includes the heliotropes (4), wild comfrey, stickseeds, for-get-me-nots, scorpion grasses. puccoons, and western false gromwell.

VERVAIN FAMILY—(*Verbenaceae*).—Includes the verbenas or vervains, of which there are 11 species; the lippia (fogfruit); and the French mulberry, a fruit with a blue, berry-like group, with 4 nutlets, but does not belong to the mulberry family.

HEATH FAMILY—(*Ericaceae*).—This family includes the Indian pipe or ghost flower, the azeleas, male berry, and the huckleberries. All the plants of this group growing are found in the eastern and southeastern parts of the State.

PRIMROSE FAMILY—(*Primulaceae*).—This family includes the water pimpernells, loosestrifes, chaffweeds, shooting star (American cowslip), Indian chief, rooster head. The plants of this group occur in moist soils and along streams in the southern and southeastern part of the State.

SAPODILLA FAMILY—(*Sapotaceae*).—This family is represented in our range by only one species, chittim-wood, the woody buckthorn (Bumelia launginosa). This species occurs in open woods throughout the State. Southern buckthorn occurs in the Arbuckle Mountains (*Bumelia lyciodesQ*.

EBONY FAMILY—(*Ebenaceae*).—This family is represented by one species in our State, the common persimmon (*Diospyros virginia*). This species occurs as a shrub or small tree in creek valleys and draws throughout the State, except in the Panhandle counties.

STORAX FAMILY—(*Slyracaceae*).—This family is represented by the silverbell or snowdrop tree. It occurs in the southeastern part of the State.

OLIVE FAMILY—(*Oleaceae*).—This family includes the ashes of which there are at least 3 species in the State; the swamp privets; and the fringe-tree or old man's beard.

MINT FAMILY—(*Labiatae*).—This family includes a large number of plants which are the fermanders (woodsage), pennyroyal, skullcaps (6), common or white horehound, hyssop, catnip or catmint, ground ivy, carpenter weed or heal-all, dragonhead or lion's heart, henbit, common motherwort, hedgenettle, sage, horsemint, woodmint, pennyroyal (3), low calamint, mountainmint, water horehounds, spearmint and peppermint.

POTATOE FAMILY—(*Solanaceae*).—This family includes the night shades (8), among which are the various night shades, named chiefly from the shape of the leaf; the horsenettle; and buffalo burr (bull-nettle, Santa Fe thistle, sand burr); the ground cherries (31); and the jimson weeds.

FIGWORT FAMILY—(*Scrophulariaceae*).—This family includes the mullins (2), found in the eastern and northeastern parts of the State, the toad flaxes (2) (butter and eggs, blue toad-flax), the figworts, beard-tongues, monkey flowers, hyssops, mullin (fox-glove), gerardia (9), blue-hearts, painted cup, and Indian paint brush.

BLADDERWORT FAMILY—(*Lentibulariaceae*).—This family includes certain low, delicate herbs living in water or moist soils. The family is represented by 4 or 5 species occuring in the eastern half of the State, and in the vicinity of the Wichita Mountains.

BROOM-RAPE FAMILY—(*Orobranchaceae*).—This family is represented by 2 species—the Louisiana broom-rape, found in sand soils in the western end of the State, and in the Panhandle counties, and the one flowered broom-rape or cancer root, growing parastic on the roots of several species of herbaceous plants in the eastern part of the State.

CATALPA FAMILY—(*Bignoniaceae*).—The family is represented in Oklahoma by the trumpet creeper vine (*tecomma radiacans*); and the western catalpa,(hardy catalpa, cigar tree, catawba tree) (*catalpa speciosa*), and the southern catalpa or Indian bean (*catalpa bignoniydes*). Both of these catalpas are planted for shade and the former one especially being extensively grown.

MARTYNIA FAMILY—(*Martyniaceae*).—This family is represented in the western part of the State by a single genus which includes one species so far as known—devil's claw or unicorn plant.

ACANTHUS FAMILY—(*Acanthaceae*).—This family includes the water-willows (3) and the ruella.

LOPSEED FAMILY—(*Phrymaceae*).—The family is represented by a single species of lopseed found in woods and thickets in the northern and northeastern parts of the State.

PLAINTAIN FAMILY—(*Plantaginaceae*).—This family includes the common plaintains (12), widely distributed over the State.

MADDER FAMILY—(*Rubiaceae*).—This family includes several species of plants within the State, among which are the bed-straws, wild liquorice, the button weeds, partridge berry or tern berry, butterbush or globe flower and the bluets.

HONEY-SUCKLE FAMILY—(*Caprifoliaceae*).—Includes the honeysuckles, found chiefly in the eastern and southern parts of the State, the coral berry or buck-bush (*symphoricarpos symphoricarpos*), in woods along streams throughout the State, except the extreme northwestern and Panhandle counties, black haws (2), and the common elder, found along streams over the entire State, except in the Panhandle counties.

VALERIAN FAMILY—(*Valerianaceae*).—This group includes 2 or 3 species among which are the cornsalids found in the eastern and southern parts of the State.

GOURD FAMILY—(*Cucurbitaceae*).—This family includes the wild gourds, squashes, and pumpkins.

THE CAMPANULA FAMILY—(*Campanulaceae*).—Herbs with milky sap, among which are the venus-looking glass, and bell flowers.

LOBELIA FAMILY—(*Lobeliaceae*).—Includes lobelias (6).

THISTLE FAMILY—(*Compositae*).—This family includes a large number of our herbs and a few of the shrubs. Among these plants are the iron weeds (veronia) (7), bonesets or thoroughworts (10) (*eupatorium*) elephant's foot (*elephantopus*), blazing stars (5), rosinweeds or gumplants (3), broom weeds, kindling weed or august flower, golden asters (4), goldenrods (21), sand daisies, daisies, asters (20), fleabanes (11), willow baccharis (in river valleys in the western half of the State), rabbit tobacco, marshfleabanes, cat's-foot, pearly everlasting, white balsam, sweet life everlasting, purple cudweed, leaf-cups, rosin-weeds (5), ferverfew, ragweeds (5), cockle-burs (3), prairie zinnia, false sunflowers, cone-flowers (14), column flowers, sunflowers (20), coreopsis (7), beggarticks or sticktights, including the Spanish needles, sneeze weeds, yarrows, dog-fennel (may-weed), feted marigold, ox-eyed daisy, sage brushes (11),

Indian Plantains, common groundsel, butterweed, ragworts, squawweed, common burdock, thistles (5), American starr thistle, chickory, dwarf dandelion (3), false dandelion, common dandelion, sow thistles, including the spring sow thistle and rabbit lettuce, wild lettuce (8), rattlesnake root, white lettuce, and hawkweeds (3).

TREES AND SHRUBS.

Far back in geologic history we find the records of tree growth.

Our coal beds are the products of abundant growth of vegetation, consisting in a large part of trees, many of which reached large proportions. Fossil leaves, plants, and twigs in the shales and sandstones, and the silicified trunks of trees commonly found bear out the extensive growth of trees in times past.

Within recent times the greater part of the land surface of the earth has been covered with forest growth. At the present time trees are among the most conspicuous objects of nature. It has been estimated that the aggregate original area of the forests of North America was about 850 million acres. This original area has been reduced until not more than 550 million acres remain, and a large percentage of this is not to be considered in the area of the economic timbered region.

A tree is usually defined as a plant with a single trunk of woody structure that does not branch for some distance above the ground. Woody plants that branch directly above the soil, although they grow to a height of 20 feet or more, are called shrubs or bushes. Certain shrubs may be made to grow treelike, and some plants which are shrubs in the northern states grow as trees of considerable size farther south.

A tree consists of three parts; first, the root; second, the branch or stem; third, the crown. The root extends into the ground from a few inches to several feet, according to the variety of tree, character of soil, and the moisture conditions. The trunk or stem supports the crown and supplies it with mineral food and water from the roots. The crown consists of a network of branches, buds, and leaves. The most important processes in the growth, reproduction and nourishment of the tree, take place in the crown.

Trees and shrubs as they stand together on some large area whose principal crop is trees and shrubs constitutes a forest. However, a forest is far more than a collection of trees. It has a population of animals and plants peculiar to itself, a soil largely of its own making, and climatic conditions differing from those in the open country. The forest prevents floods, influences rainfall, supplies fuel, timber, and food.

But with all the good that the forest has been to man its real value has not been considered and it has been recklessly used and ruthlessly destroyed.

The United States is divided into several forest regions. A single unbroken forest belt—the spruce forest of Canada, extends across North America. *(1) A part of this belt extends within the borders of the United

(1) Bowman, Isaac, Forest Physiography, p. 123.

States in the region about the Great Lakes. This region has the typical growth of black and white spruce, poplar, canoe birch, aspen, and tamarack.

To the southward from the south limit of this spruce forest are marked forest regions extending in a north-south line. These are the Atlantic forest, the Pacific forest, the Rocky Mountain forest, and the forested regions of the Great Basin. The Great Plains region is forestless but not treeless.

Oklahoma is in part in the forest region of the Great Basin and part in the region of the Great Plains. The east one-third of the State lies chiefly in the timbered belt. The southeastern part of the State has a forest growth, which places it in the southern forest belt which is chiefly characterized by the long leaf, short leaf, loblolly, and slash pine. In the northeastern part of the State east of Grand River and north of the Arkansas is found a tree growth very characteristic of a hardwood forest of the central Mississippi valley. The remainder of the principal timbered region of the State lies within the Sandstone Hills region, except the area in the Arbuckle Mountain and the Cretaceous plains along Red River.

A strip 75 miles wide clear across the east side of the State is covered with heavy timber. The northern part of the strip is hard wood. South of Arkansas River it is chiefly pine, hickory, and oak. There are millions and millions of feet of valuable timber. The pine is of the yellow variety. The hardwood is oak, hickory, walnut, elm, and maple. In the southeastern part of the State pine timber is of much more value than the land on which it stands. One lumber company, it is said, owns enough timber to keep its mills running 20 years if they cut 100,000 feet every day.

In the western half of the State trees are found usually only along the streams or on the sandhills. The level uplands are grass covered. Cottonwood, elm, hackberry, chinaberry, walnut, willow, dogwood, redbud, soapberry, box elder and mulberry are the chief kinds found. On the flat prairies and along some of the streams in the western part is found the thorny mesquite. Blackjack, barren, post oak, hackberry, hickory, and sumac grow on the sand hills. On the high plains in the extreme western part of the State there are very few trees even along the streams, but even here an occasional low elm or dwarf cottonwood or willow may be found. One of the problems which confronts the citizens is to discover what trees may be transplanted and grown in the treeless part of the State. The inhabitants desire shade trees and it is important that trees be secured which will meet the conditions. Difficulty will often be experienced in getting trees started in this region, but many efforts will meet with success.

The black locust, soft maple or silver maple, box elder or ash leaved maple, the green ash, mulberry, and catalpa are being most widely planted for shade in our cities. Cottonwood may also be added to the list. The chief reason for selecting most of these trees is that they are quick growing. The locust if used will have to be safeguarded against the locust borers which have wrought such great havoc in other states; the catalpa against the catalpa sphinx. The cottonwood, while a very rapid grower, has its faults. It sheds leaves continually through the summer and becomes bare in early fall, and the roots find their way into wells and sewers and do much damage. In many eastern cities ordinances have been passed requiring all cottonwoods within 100 feet of sewers to be cut. The walnut, red bud and oaks may also be added to the list for shade. In Oklahoma City many oaks and cottonwoods of the original forest still stand for shade. Among imported trees suitable for street and lawn planting are the dwarf catalpa, linden, tree of Heaven, Paulounia, and Koelreuteria paniculata, a native of China. Fruit trees may also be used for shade even for planting along streets and roadways. People will soon learn to respect the trees and they will not be abused in securing the fruit.

Trees have two distinct modes of branching—one has a central stem running the entire length of the tree as in the hickory or pine. This is called the excurrent growth. In the second class the main stem is soon lost in the branches as in the walnut or catalpa. This is known as delinquescent growth. Two kinds of roots are found on trees—those having a tap root as in the hickory and oaks. These trees are difficult to transplant. Those having multiple roots in which the roots are numerous and extend outward from the base, as in the maples and elms. Such trees are easily transplanted.

Willows and poplars may be started by branches placed in the ground. Most trees are best grown from the seed. Trees best for transplanting are those not more than two years old.

A LIST OF TREES AND SHRUBS OF OKLHAMOA.

Trees and shrubs are plants which belong to the sub-kingdom—SPERMATOPHYTA—the seed-bearing plants. The plants of this division belong to two classes—the GYMNOSPERMAE, plants with naked seed; and ANGIOSPERMAE, plants with protected seeds.

CLASS I. GYMNOSPERMAE.

Naked Seeds.

The Gymnosperms are an ancient group of plants. They are found among the fossils of the Silurian age, and are most numerous in Triassic time. They are now represented by about 450 species of trees and shrubs. They are flowering plants in which the ovules, or seeds, are not inclosed in an ovary, but are borne naked upon an open scale. This scale is a modified leaf. The plants have a resinous juice, and chiefly parallel-veined, needle-shaped, or scale-like evergreen leaves. By far the larger number of the Gymnosperms are Coniferous (Pinacae or Coniferae including some shrubs) or cone-bearing trees, the fruit being called a cone because of its form. This family of the class is the only one represented in our area.

PINE FAMILY. PINACEAE.

Pines. Conifers.
Pinaceae. Coniferae.

The trees belonging to this family are commonly known as Evergreens, because with the exception of the Larch and the Bald Cypress they do not shed their leaves during the winter.

PINES. GENUS PINUS.

Loblolly pine. (Old Field pine. Rosemary pine).
Pinus Taedea.

Yellow pine. (Short-leaf pine. North Carolina pine, Spruce pine.)
Pinus echinata.

Southern or Bald cypress. (Deciduous cypress.)
Taxodium distichum.

Red cedar. (Common cedar. Juniper.)
Juniperus Virginiana.

Juniper. (Ground cedar.)
Juniperus communis.

Arbor vitae. (White cedar.)
Thuja occidentalis.

CLASS II. ANGIOSPERMAE.

Enclosed Seeds.

MAGNOLIA FAMILY. MAGNOLIACEAE.

Found in the extreme eastern and southeastern part of the State.

TULIP TREE. GENUS LIRIODENDRON.

Tulip tree. (Yellow poplar. Tulip poplar.)
Liriodendron tulipifera.

CUSTARD APPLE FAMILY. ANONACEAE.

PAPAWS. GENUS ASIMINA.

Common papaw. (Found only in the eastern part of the State)
Asimina triloba.

TAMARISK FAMILY. TAMARISCINEAE.

The *Tamarisks* are sub-evergreen, shrubs or small trees, with very small pinkish flowers, in spike-like clusters, or thickly grouped along the slender drooping branches. The leaves are very small and scale-like. There are several species in cultivation, and numerous names have been applied by nursery men. Many of the members of this family are found growing escaped from cultivation about old building places. Some of the varieties are now being used to considerable extent for hedges about lawns and gardens.

MALLOW FAMILY. MALVACEAE.

The *Hibiscus* or *Rose Mallows* comprise a large family of mainly herbaceous plants with large holly-hock-like flowers. There is but one cultivated species in this region which has woody structure and tree-like appearance. This is the *Shrubby Althea* or *Hibiscus Syriacus* commonly called *Tree Hibiscus* or *Rose of Sharon.* This was originally introduced from Syria. There are many varieties with single or double flowers of many colors,—white, pink, red, purple, and variegated. A large number of these have been planted for ornament over the State and some have been observed which had reached a height of 15 to 20 feet.

LINDEN FAMILY. TILIACEAE.

LINDENS OR BASSWOOD. GENUS TILIA.

Basswood. (Whitewood. Linden. Lime Tree.)
Tilia Americana.

The range of the basswood extends through the eastern half of Oklahoma and though it is not abundant, is found scattering here and there in the areas of good soil. It is also in cultivation in the State. Another tree, probably a variety of the above, is found in cultivation. This is *Tilia pubescens,* small leave basswood. Several introduced species and nursery varieties are much valued for their dense foliage, odd-shaped leaves, and slender branches.

RUE FAMILY. RUTACEAE.

PRICKLY ASHES. GENUS XANTHOXYLUM.

Prickly ash. (Toothache tree.)
Xanthoxylum Clava-Hercules.

WAFER ASHES. GENUS PTELEA.

Wafer ash. (Hop-tree. Shrubby trefoil.)
Ptelea trifoliata.

This shrub, often growing with the true shape of a tree, never attains a height of more than a few feet (6-10). The fruit resembles that of the elm but is larger. It is known to occur from the east side to as far west as Alva in Woods County.

CORK TREE. GENUS PHELLODENDRON.

Chinese cork tree.
Phellodendron Amurense.

An introduced tree with compound leaves resembling those of the Tree of Heaven. The cork tree is being planted in some of our city parks, and grows fairly well.

MELIA FAMILY. MELIACEAE.

CHINA TREE. GENUS MELIA.

China-tree. (China-Berry tree. Pride of India.)
Melia Azedarach.

QUASSIA FAMILY. SIMARUBACEAE.

TREE OF HEAVEN. GENUS AILANTHUS.

Tree of Heaven. (Paradise tree. Chinese sumac.)
Ailanthus glandulosus.

This tree was originally introduced from China and Japan, but it is now widely cultivated and naturalized. It is found growing wild in many places, having escaped from cultivation. It is a very rapid-growing tree and hardy throughout. At all times of the year it has a good appearance and is well adapted for shade and ornamental purposes.

HOLLY FAMILY. ILICINAE.

HOLLIES. GENUS ILEX.

American holly. (Christmas holly.)
Ilex apaca.

Swamp holly. (Deciduous holly.)
Ilex decidua.

Ink berry.
Ilex glabra.

STAFF-TREE FAMILY. CELASTRACEAE.

WAHOO. GENUS EUONYMUS.

Wahoo. (Burning Bush. Spindle tree.)
Euonymus atropurpureus.

BITTER SWEET. GENUS CELASTRUS.

Bitter sweet. (Waxwork.)
Celastrus scandens.

SOAPBERRY FAMILY. SAPINDACEAE.

SOAPBERRY. GENUS SAPINDUS.

Soapberry (c). (Western soapberry. Wild China tree.)
Sapindus Drummondi.

A small tree common throughout the State, except extreme northwest. Rare in some localities. Easily distinguished by its clusters of fruit remaining more or less shriveled on the tree until spring. The largest tree observed was in the Grand River Valley and was 10 inches in diameter and more than 30 feet high. Often a large number of the trees are found growing together.

HORSE-CHESTNUT FAMILY. HIPPOCASTANACEAE.

HORSE CHESTNUTS. GENUS AESCULUS.

Common horse chestnut.
Aesculus Hippocastanum.

Yellow buckeye. (Sweet buckeye.)
Aesculus octandra.

Ohio buckeye. (Sweet buckeye. Fetid buckeye.)
Aesculus glabra.

GENUS KOELREUTERIA.

Koelreuteria paniculata.

MAPLE FAMILY. ACERACEAE.

MAPLES. GENUS ACER.

Silver maple. (Soft maple. White maple. River maple.)
Acer saccharinum.

Red, scarlet, or water maple. (Swamp maple.)
Acer rubrum.

Hard maple. (Sugar or rock maple. Sugar-tree.)
Acer saccharum.

Black maple. (Sugar maple.)
Acer nigrum.

Box Elder. (Ash-leaved maple.)
Acer negundo.

The NORWAY MAPLE, *acer platanoides,* and CUT-LEAVED JAPANESE MAPLE, *acer palmatum,* are in cultivation, but so far have not made very successful growth in the localities where planted.

BLADDER-NUT FAMILY. STARPHYLEACEAE.

American Bladder-nut.
Staphylea trifoliata.

CASHEW FAMILY. ANACARDIACEAE.

SUMACHS. GENUS RHUS.

Smooth sumac. (Upland or scarlet sumac.)
Rhus glabra.

Dwarf, black or mountain sumac. (Upland sumac.)
Rhus copallina.

Poison sumac. (Poison dogwood. Poison elder.)
Rhus Venenata.

Cut-leaved sumac.
Rhus lacinata.

Ill scented sumac. (Skunk bush.)
Rhus trilobata.

Fragrant or sweet-scented sumac.
Rhus aromatica.

Poison ivy. (Poison oak. Climath. Three leaf ivy.)
Rhus radicans.

Wild or American smoke tree. (Chittam wood.)
Rhus cotinoides.

PULSE OR PEA FAMILY. LEGUMINOSAE.

LOCUSTS. GENUS ROBINIA.

Common locust. (Yellow locust. Black locust. Silver chain.)
Robinia pseudacacia

REDBUDS. GENUS CERCIS.

Judas tree. (Redbud.)
Cercis canadensis.

Common throughout except far western part of State. Grows as a shrub to small tree, reaches height of 20 feet. Excellent for lawn planting.

COFFEE TREE. GENUS GYMNOCLADUS.

Kentucky coffee tree. (Coffee-nut.)
Gymnocladus Canadensis.

HONEY-LOCUSTS. GENUS GLEDITSCHIA.

Honey locust.
Gleditschia triacanthos.

Water locust.
Gleditschia aquatica.

ROSE FAMILY. ROSACEAE.

PEACHES, PLUMS, AND CHERRIES. GENUS PRUNUS.

Common peach.
Prunus Persica.

Nectarine. (Smooth-skinned peach.)
Prunus laevis.

Pigeon cherry. (Pincherry. Bird cherry. Wild red cherry.)
Prunus Pennsylvanica.

Wild black cherry. (Cabinet or river cherry.)
Prunus serolina.

American plum. (Wild plum.)
Prunus Americana.

Wild goose plum. (River plum.)
Prunus hortulana.

Chickasaw plum.
Prunus Augustifolia.

Watson's plum. (Sand plum.)
Prunus Watsoni.

Low plum.
Prunus gracilis.

Choke cherry.
Prunus Virginiana.

Western sand cherry. (Bessy's cherry.)
Prunus Besseyi.

APPLES AND PEARS. GENUS PYRUS.

Common apple
Pyrus malus.

Soulard crab.
Pyrus soulardi.

Prairie crab.
Pyrus isensis.

Common pear.
Pyrus communis.

American or garland crab. (Fragrant crab.)
Pyrus coronaria.

American or small fruited ash.
Pyrus americana.

Elder-leaved or large fruited mountain ash.
Pyrus sambricifolia.

GENUS AMELANCHIER.

June-berry. (Service-berry. Shad bush.)
Ameleanchier Canadensis.

HAWS OR THORNS. GENUS CRATAEGUS.

Green haw.
Crataegus viridis.

Washington thorn.
Crataegus cordata.

Dotted fruited hawthorn.
Crataegus punctata.

GENUS RUBUS.

WITCH HAZEL FAMILY. HAMAMELIDACEAE.

GENUS HAMAMELIS.

Witch hazel.
Hamamelis Virginiana.

Found in central-east side, north of Arkansas River. The largest observed was south of Marble City.

GENUS LIQUIDAMBAR.

Sweet gum. (Bilsted. Red gum.)
Liquidambar styraciflua.

This tree is found in the east side of the State south of Arkansas River. It is abundant along Poteau and Kiamichi rivers and their tributaries. It is a tree worthy of extensive cultivation.

GINSENG FAMILY. ARALIACEAE.

GENUS ARALIA.

Angelica tree. (Hercules club.)
Aralia spinosa.

The range of this tree would include the southeastern corner of the State; however, so far as known it has not been found. It is used for ornamental planting and is worthy of extensive use.

DOGWOOD FAMILY. CORNACEAE.

DOGWOOD OR CORNELS. GENUS CORNUS.

Flowering dogwood.
Cornus florida.

Rough leaf or white fruited dogwood.
Cornus asperifolia.

TUPELOS. GENUS NYSSA.

Pepperidge. (Black or sour gum.)
Nyssa sylvatica.

HONEY-SUCKLE FAMILY. CAPRIFOLIACEAE.

NANNY-BERRIES. GENUS VIBURNUM.

Rusty nanny berry. (Southern nanny berry.)
Viburnum rufidulum.

Small black haw.
Viburnum globosum.

Cranberry tree. (Snow ball or Guelder rose.)
Viburnum opulus.

GENUS SYMPHORICARPOS.

Coral-berry.
Symphoricarpos symphoricarpos.

Wolfberry.
Symphoricarpos occidentalis.

GENUS LONICERA.

Tartarian bush honeysuckle.
Lonicera Tartarica.

Honeysuckle.
Lonicera.

GENUS DIERVILLA.

Bush honeysuckle.
Diervilla diervilla.

GENUS SAMBUCUS.

American elder. (Sweet elder. Elderberry.)
Sambucus Canadensis.

SAPODILLA FAMILY. SAPOTACEAE.

BUMELIAS. GENUS BUMELIA.

Wooly bumelia. (Chittim wood. Buckthorn. Gum elastic.)
Bumelia languinosa, also *Bumelia lycioides.*

BUCKTHORN FAMILY. RHAMNACEAE.

BUCKTHORNS. GENUS RHAMNUS.

Yellow buckthorn. (Indian cherry.)
Rhamnus Caroliniana.

GENUS CEANOTHUS.

New Jersey tea. (Red root.)
Ceanothus Americanus.

EBONY FAMILY. EBENACEAE.

PERSIMMONS. GENUS DIOSPYROS.

Common persimmon.
Diospyros Virginiana.

Japan persimmon. (c).
Diospyros Kaki.

STORAX FAMILY. STYRACACEAE.

SILVER BELL TREES. GENUS MOHRODENDRON.

Silver bell tree. (Snow drop tree.)
Mohrodendron Carolinum.

OLIVE FAMILY. OLEACEAE.

ASHES. GENUS FRAXINUS.

White ash.
Fraxinus Americana.

Swamp ash or water ash.
Fraxinus platycarpa.

Blue ash.
Fraxinus quadrangulata

Green ash.
Fraxinus viridis.

FORESTIERA. GENUS ADELIA.

Adelia. (Forestiera. Swamp privet.)
Adelia acuminata.

PRIVET. GENUS LIGUSTRUM.

Privet. (Prim.)
Ligustrum Vulgari.

SYRINGIA. GENUS SYRINGIA.

Common lilac.
Syringia vulgaris.

Japan lilac.
Syringia Japonica.

GENUS CHIONANTHUS.

Fringe tree.
Chionanthus Virginica.

FIGWORT FAMILY. SCROPHULARIACEAE.

GENUS PAULOWNIA.

Imperial paulownia. (Paulownia tree.)

BIGNONIA FAMILY. BIGNONIACAE.

CATALPAS. GENUS CATALPA.

Indian bean. (Southern catalpa.)
Catalpa bignonioides or *Catalpa catalpa.*

Hardy catalpa. (Western catalpa. Catawba tree.)
Catalpa speciosa.

Japanese catalpas. (c).
Catalpa Kaempferi and *Catalpa Bungei.*

Dwarf forms growing from 5 to 10 feet, umbrella-shaped tops. Grown in cities and parks.

LAUREL FAMILY. LURACEAE.

BAY TREE. GENUS PERSEA.

Red bay.
Persea Borbonia.

SASSAFRAS. GENUS SASSAFRAS.

Sassafras.
Sassafras sassafras or *Sassafras officinale.*

GENUS LINDERA.

Spice bush. (Benjamin bush.)
Lindera Benzoin.

OLEASTER FAMILY. ELAEAGNACEAE.

Several members of the family in cultivation. A specimen of the RUSSIAN OLIVE, or OLEASTER, was observed which was at least 20 feet in height. The DESERT WILLOW, *chilopsis linearis,* from Texas is found in Oklahoma City. The BUFFALO BERRY or RABBIT BERRY has been found by the writer growing along the Canadian in the western part of Cleveland County.

NETTLE FAMILY. URTICACEAE.

ELMS. GENUS ULMUS.

American or white elm.
Ulmus Americana.

Slippery or red elm.
Ulmus fulva or *U. pubescens.*

Wahoo or winged elm.
Ulmus alata.

GENUS PLANERA.

Planer tree. (Water elm.)
Planera aquatica.

HACKBERRIES. GENUS CELTIS.

Sugar berry. (Hackberry.)
Celtis occidentalis.

Hackberry.
Celtis var.

Mississippi hackberry. (Sugar hackberry.)
Celtis Mississippiensis.

OSAGE ORANGE. GENUS MACLURA.

Osage orange. (Bow-wood. Bois d'Arc.)
Toxylon pomiferum, or Maclüra aurantiaca.

MULBERRIES. GENUS MORUS.

Red mulberry.
Morus rubra.

White mulberry.
Morus alba.

Both of these mulberries are grown for shade. The largest white mulberry known in the State is in Norman. This tree is a foot in diameter and 50 feet high. The Russian or barren mulberry, extensively grown for shade, is the staminate white mulberry. It is a rapid grower and a good tree to plant.

GENUS BROUSSONETIA.

Paper mulberry.
Broussonetia papyrifera.

French mulberry.
Callicarpa Americana.

A shrub reported as occuring in the northeast corner of the State. Does not belong to the Mulberry family.

PLAN-TREE FAMILY. PLATANACEAE.

GENUS PLATANUS.

Sycamore. (Buttonwood.)
Platanus occidentalis.

Oriental plane. (c).
Platanus orientalis.

WALNUT FAMILY. JUGLANDACEAE.

WALNUTS. GENUS JUGLANS.

Butternut. (White walnut.)
Juglans cinerea.

Black walnut.
Juglans nigra.

HICKORIES. GENUS CARYA OR HICORIA.

Bitternut hickory. (Swamp hickory.)
Hickory minima.

Water hickory. (Swamp hickory.)
Hicoria aquatica.

Big shell bark hickory. (King nut.)
Hicoria ovata or *Carya sulcata.*

Mocker-nut hickory.
Hicoria alba or *Carya tomentosa.*

Pignut hickory.
Hicoria glabra or *Carya porcina.*

Pecans.
Hicoria pecan.

Pale-leaf hickory.
Hicoria villosa.

WILLOW FAMILY. SALICACEAE.

WILLOWS. GENUS SALIX.

Peach or almond-leaf willow. (Western black willow.)
Salix amygdaloides.

Long-stalked willow. (Ward willow.)
Salix longipes.

Black willow.
Salix nigra.

Narrow or long-leaved willow. (Sand-bar willow. Ring willow.)
Salix longifolia or *S. fluviatilis.*

POPLARS. GENUS POPULUS.

White poplar. (Abele tree. Silver-leaf poplar.)

Populus alba.

Cottonwood. (Necklace poplar. Carolina poplar.)
Populus deltoides.

Western cottonwood.
Populus deltoides occidentalis.

Lombardy poplar.
Populus nigra Italica or *P. Dilatata.*

Swamp poplar. (Downy-leaved poplar.)
Populus heterophylla.

Lance-leaf cottonwood and narrow-leaf cottonwood.
Populus acuminata. *Populus angustifolia.*

OAK FAMILY. CUPULIFERAE.

BIRCHES. GENUS BETULA.

River birch. (Water birch. Red birch.)
Betula nigra.

European white birch.
Betula alba.

ALDERS. GENUS ALNUS.

Sea side alder.
Alnus maratima.

Smooth alder.
Alnus rugosa or *A. serrulata.*

HAZELNUTS. GENUS CORYLUS.

Wild hazlenut or common hazlenut.
Corylus Americana.

HORNBEAMS. GENUS CARPINUS.

American hornbeam. (Blue or water beech.)
Carpinus Caroliniana.

HOP HORNBEAM. GENUS OSTRYA.

Iron-wood. (American hop-hornbeam.)
Ostrya Virginiana.

OAKS. GENUS QUERCUS.

Red oak.
Quercus rubra.

Southern red oak. (Schneck's oak.)
Quercus Texana.

Pin oak. (Swamp oak.)
Quercus palustris.

Spanish oak.
Quercus digitata, or *Quercus falcata.*

Yellow oak. (Black oak. Quercitron oak.)
Quercus velutina.

Black jack oak. (Barren oak.)
Quercus Marilandica or *Quercus nigra.*

Water oak. (Duck oak. Possum oak.)
Quercus nigra or *Quercus aquatica.*

Willow oak.
Quercus phellos.

Shingle oak.
Quercus imbricaria.

White oak.
Quercus alba.

Post oak. (Rough or box white oak.)
Quercus minor or *Q. obtusiloba* or *Q. stellata.*

Burr oak. (Mossy cup oak.)
Quercus macrocarpa.

Over-cup oak.
Quercus lyrata.

Swamp white oak.
Quercus platanoides.

Cow oak. (Basket oak.)
Quercus Michauxii.

Chestnut oak. (Chinquapin oak. Yellow oak.)
Quercus acuminata or *Q. muhlenbergii.*

Live oak.
Quercus Virginiana.

Dwarf chestnut oak. (Scrub oak.)
Quercus prinoides.

CHESTNUTS. GENUS CASTANEA.

Chestnut.
Castanea sativa or *C. dentata.*

Chinquapin.
Castanea pumila.

BEECHES. GENUS FAGUS.

American beech.
Fagus grandifolia.

HUCKLEBERRY FAMILY. VACCINICAEAE.

GENUS BATODENDRON.

Farkleberry. (Tree huckleberry. Sparkleberry.)
Batodendron arboreum.

MADDER FAMILY. RUBIACEAE.

GENUS CEPHALANTHUS.

Button bush.
Cephalanthus occidentalis.

HYDRANGA FAMILY. HYDRANGEACEAE.

GENUS HYDRANGEA.

Wild hydrangea.
Hydrangea arborescens.

GENUS PHILADELPHUS.

Syringa.
Philadelphus.

GOOSEBERRY FAMILY. GROSSULARIACEAE.

GENUS RIBES.

Gooseberries.
Ribes.

Common currants. (c).
Ribes rubrum and *ribes nigrum.*

Slender gooseberry.
Ribes gracile.

Missouri gooseberry.
Ribes Missouriensis.

Golden buffalo or Missouri currant. (Clove bush.)
Ribes aureum chrysococcum.

MISCELLANEOUS.

Prairie mesquite.
Prosopis glandulosa.

False or bastard indigo.
Amorpha fruticosa.

Lead plant. (Shoestrings.)
Amorpha canescens.

Trumpet flower or trumpet creeper.
Tecoma radicans.

American mistletoe.
Phoradendron flavescens.

Made in the USA
Monee, IL
18 October 2026

66036442R00049